The Minitab Manual

**Dorothy Wakefield
Kathleen McLaughlin**

Elementary Statistics
PICTURING THE WORLD

LARSON ■ FARBER

PRENTICE HALL, Upper Saddle River, NJ 07458

Acquisitions Editor: Kathy Boothby Sestak
Supplement Editor: Joanne Wendelken
Special Projects Manager: Barbara A. Murray
Production Editor: Shea Oakley
Supplement Cover Manager: Paul Gourhan
Supplement Cover Designer: Liz Nemeth
Manufacturing Buyer: Alan Fischer

© 2000 by Prentice Hall
Upper Saddle River, NJ 07458

All rights reserved. No part of this book may be reproduced, in any form or by any means, without permission in writing from the publisher

Printed in the United States of America

10 9 8 7 6 5 4 3 2 1

ISBN 0-13-015210-2

Prentice-Hall International (UK) Limited, London
Prentice-Hall of Australia Pty. Limited, Sydney
Prentice-Hall Canada, Inc., Toronto
Prentice-Hall Hispanoamericana, S.A., Mexico
Prentice-Hall of India Private Limited, New Delhi
Prentice-Hall (Singapore) Pte. Ltd.
Prentice-Hall of Japan, Inc., Tokyo
Editora Prentice-Hall do Brazil, Ltda., Rio de Janeiro

▶ Introduction

The MINITAB Manual is one of a series of companion technology manuals that provide hands-on technology assistance to users of Larson/Farber *Elementary Statistics: Picturing the World.*

Detailed instructions for working selected examples, exercises, and Technology Labs from *Elementary Statistics: Picturing the World* are provided in this manual. To make the correlation with the text as seamless as possible, the table of contents includes page references for both the Larson/Farber text and this manual.

All of the data sets referenced in this manual are found on the data disk packaged in the back of every new copy of Larson/Farber *Elementary Statistics: Picturing the World.* If needed, the MINITAB files (.mtp) may also be downloaded from the texts' companion website at www.prenhall.com/Larson.

▶ Contents:

	Larson/Farber Elementary Statistics: Page:	The MINITAB Manual Page:
Getting Started with MINITAB		1
Chapter 1 Introduction to Statistics		
Technology Lab *Generating Random Numbers*	22-23	8
Chapter 2 Descriptive Statistics		
Section 2.1:		
Example 7 *Constructing a Histogram*	38	17
Exercise 17	41	23
Exercise 21	42	26
Exercises 31a	43	28
Section 2.2		
Example 2 *Constructing a Stem-and-Leaf Plot*	45	29
Example 3 *Constructing a Dot Plot*	46	31
Example 4 *Constructing a Pie Chart*	47	33
Example 5 *Constructing Pareto Charts*	48	35
Example 7 *Constructing a Time Series Chart*	50	38
Exercise 17	52	39
Exercise 21	53	40
Exercise 23	53	41
Exercise 25	53	42
Exercise 27	54	43
Exercise 29	54	44
Section 2.3		
Example 6 *Finding the Mean and Standard Deviation*	58	45
Exercise 17	63	49
Exercise 39	66	50
Exercise 45	67	51
Section 2.4		
Example 5 *Calculating the Mean and Standard Deviation*	72	52
Exercise 11	79	53
Exercise 17	79	54
Section 2.5		
Example 2 *Finding Quartiles*	86	55
Example 4 *Constructing a Box-and-Whisker Plot*	88	56
Exercise 11	90	58
Exercise 19	91	59
Technology Lab *Using Descriptive Statistics*	93	60

	Larson/Farber Elementary Statistics: Page:	The MINITAB Manual Page:

Chapter 3 Probability
Section 3.1
   Law of Large Numbers
Section 3.2
   Exercise 25
   Technology Lab *Composing Mozart Variations*
	108
	
	123
	143
		62
		
		63
		64

Chapter 4 Discrete Probability Distributions
Section 4.2
   Example 5 *Finding Probabilities*
   Example 7 *Constructing / Graphing Binomial Distribution*
Section 4.3
   Example 3 *Finding Poisson Probabilities*
   Technology Lab *Poisson Distributions as Queuing Models*
	169
	171
	
	180
	185
		66
		68
		
		70
		72

Chapter 5 Normal Probability Distributions
Section 5.1
   Exercise 27
Section 5.3
   Example 4 *Finding Normal Probabilities*
   Example 5 *Finding a specific data value*
   Exercise 3
Section 5.4
   Example 6 *Finding Probabilities for X and $\overline{X}$*
   Technology Lab *Age Distribution In the United States*
	201
	
	215
	216
	217
	
	229
	234
		74
		
		77
		79
		80
		
		81
		82

Chapter 6 Confidence Intervals
Section 6.1
   Example 4 *Construct a 99% Confidence Interval*
   Exercise 31
   Exercise 43
Section 6.2
   Example 2 *Construct a 95% Confidence Interval*
   Exercise 19
Section 6.3
   Example 2 *Construct a 95% Confidence Interval for p*
   Exercise 11
   Technology Lab *Most Admired Polls*
	256
	261
	263
	
	268
	272
	
	277
	280
	283
		85
		87
		88
		
		89
		90
		
		91
		93
		94

	Larson/Farber Elementary Statistics: Page:	The MINITAB Manual Page:
Chapter 7 Hypothesis Testing with One Sample		
Section 7.2		
Example 9 *Hypothesis Testing using P-values*	323	96
Exercise 25	326	98
Exercise 33	328	99
Section 7.3		
Example 4 *Testing μ with a Small Sample*	333	100
Exercise 23	338	101
Exercise 24	338	102
Section 7.4		
Example 2 *Hypothesis Testing for a Proportion*	342	103
Exercise 9	344	105
Exercise 11	345	106
Technology Lab *The Case of the Vanishing Woman*	346	107
Section 7.5		
Extending the Basics	355	108
Chapter 8 Hypothesis Testing with Two Samples		
Section 8.1		
Exercise 13	375	109
Exercise 15	375	111
Section 8.2		
Example 1 *Snow Tire Performance*	382	112
Exercise 17	386	113
Exercise 20	387	114
Section 8.3		
Example 2 *Golf Scores*	392	115
Exercise 16	396	117
Exercise 17	397	118
Section 8.4		
Example 1 *Difference between internet users*	402	119
Exercise 7	404	121
Exercise 10	405	122
Technology Lab *Tails over Heads*	407	123

	Larson/Farber Elementary Statistics: Page:	The MINITAB Manual Page:
Chapter 9 Correlation and Regression		
Section 9.1		
Example 3 *Constructing a Scatterplot*	420	124
Example 5 *Finding the Correlation Coefficient*	423	126
Exercise 13	428	128
Section 9.2		
Example 2 *Finding a Regression Equation*	434	130
Exercise 11	437	132
Section 9.3		
Example 2 *Finding the Standard Error and Coefficient of Determination*	445	134
Example 3 *Constructing a Prediction Interval*	447	135
Exercise 11	449	137
Exercise 19	450	137
Section 9.4		
Example 1 *Finding a Multiple Regression Equation*	452	138
Exercise 5	456	139
Technology Lab *Tar, Nicotine, and Carbon Monoxide*	457	140
Chapter 10 Chi-Square Tests and the F-Distribution		
Section 10.1		
Example 3 *The Chi-Square Goodness of Fit Test*	470	141
Exercise 5	473	144
Section 10.2		
Example 3 *Chi-Square Independence Test*	481	146
Exercise 5	483	148
Section 10.3		
Picturing the World *A Two-Sample F-Test*	491	149
Section 10.4		
Example 2 *ANOVA Tests*	501	152
Exercise 1	503	153
Exercise 5	504	154
Technology Lab *Crash Tests*	508	155

	Larson/Farber Elementary Statistics: Page:	The MINITAB Manual Page:
Chapter 11 Nonparametric Tests		
Section 11.1		
Example 3 *Using the Paired-Sample Sign Test*	521	157
Exercise 3	522	159
Exercise 15	524	161
Section 11.2		
Example 1 *Performing a Wilcoxon Signed-Rank Test*	528	162
Example 2 *Performing a Wilcoxon Rank Sum Test*	531	164
Exercise 5	534	166
Section 11.3		
Example 1 *Performing a Kruskal-Wallis Test*	539	167
Exercise 1	541	170
Section 11.4		
Example 1 *The Spearman Rank Correlation Coefficient*	544	171
Exercise 3	546	174
Technology Lab *Air Conditioner Performance and Price*	549	175

Getting Started with MINITAB

> ▶ Using MINITAB Files

MINITAB is a Windows-based Statistical software package. It is very easy to use, and can perform many statistical analyses. When you first open MINITAB, the screen is divided into two parts. The top half is called the Session Window. The results of the statistical analyses are often displayed in the Session Window. The bottom half of the screen is the Data Window. It is called a Worksheet and will contain the data.

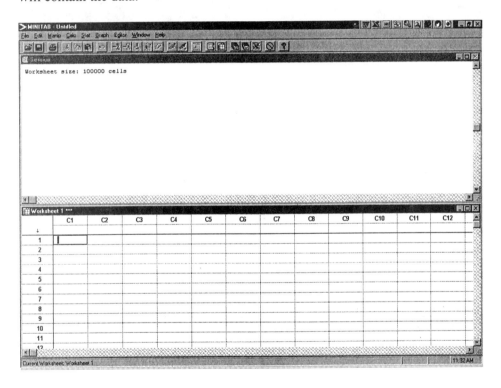

The data will either be entered directly into the Worksheet, or saved worksheets can be opened and used.

▶ Entering Data into the Data Window

To enter the data into the Data Window, you must first click on the bottom half of the screen to make the Data Window active. You can tell which half of the screen is active by the blue bar going across the screen. In the previous picture, notice that the blue bar is in the middle of the screen, highlighting **Worksheet 1**. This indicates that the Data Window is active. The bar will be gray if the Window is not active. (Notice the Session Window bar is gray.)

In MINITAB, the columns are referred to as C1, C2, etc. Notice that there is an empty cell directly below each heading C1, C2, etc. This cell is for a column name. Column names are optional because you can refer to a column as C1 or C2, but a name helps to describe the data contained in a column. Enter the data beginning in cell 1. Notice that the cell numbers are located in the leftmost column of the worksheet.

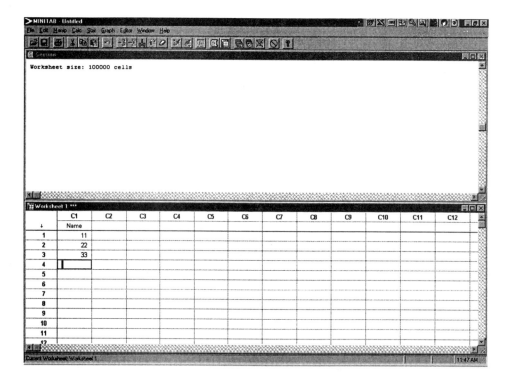

▸ Opening Saved Worksheets

Many of the worksheets that you will be using are saved on the enclosed data disk. To open a saved worksheet, click on **File → Open Worksheet.** The following screen will appear.

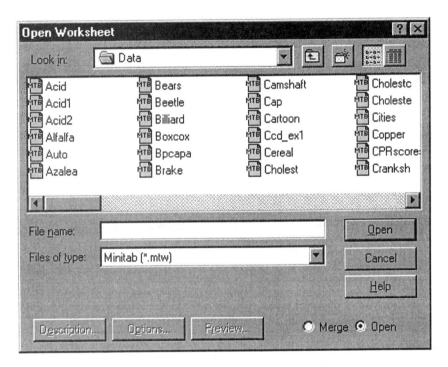

First, you must tell MINITAB where the data files are located. Since the data files are located on the data disk, you must tell MINITAB to **Look In** the 3 1/2" Floppy (A:). To do this, click on the down arrow to the right of the top input field and select the floppy drive by double-clicking on it.

4 Getting Started with MINITAB

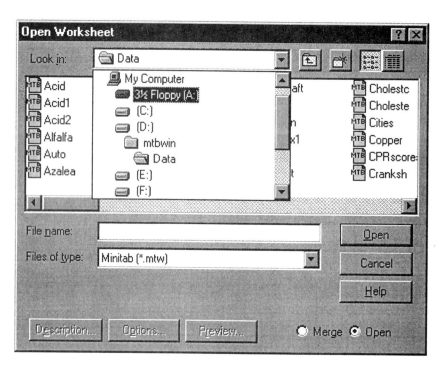

When you do this, you should see three folders listed. Select the MINITAB folder with a double-click. Now you should see a folder for each of the eleven chapters of the book.

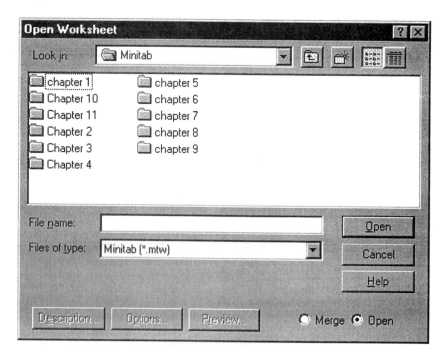

All data files are saved as MINITAB Portable worksheets and have the extension **.mtp.** Click on the down arrow for the field called **Files of type** and select **Minitab Portable (*.mtp).**

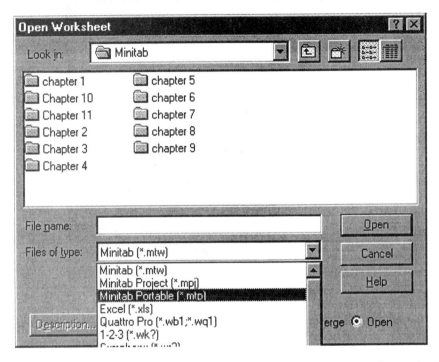

Now, select the folder called **chapter 1** (by double-clicking) and you should see all the MINITAB worksheets for Chapter 1.

6   Getting Started with MINITAB

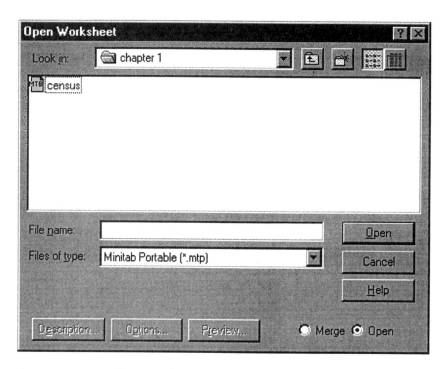

As you can see, Chapter 1 has only one worksheet saved to disk. To open the worksheet **census**, double-click on it and the worksheet should appear in the Data Window.

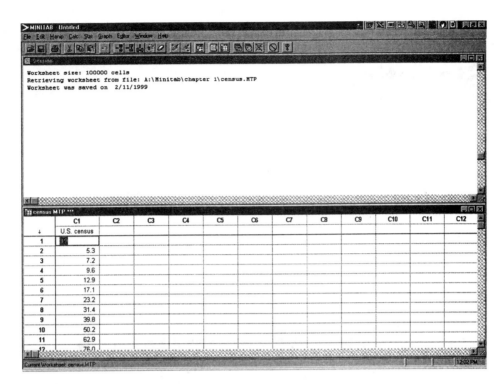

You are now ready to begin analyzing the data and learning more about MINITAB.

Introduction to Statistics

▸ **Technology Lab (pg. 22-23)** Generating Random Numbers

1. To select 8 numbers randomly from the numbers 1 to 74, first store the numbers 1 to 74 in C1. Click on **Calc → Make Patterned Data → Simple Set of Numbers.** You should **Store patterned data in** C1. The numbers will begin **From the first value** 1 and go **To last value** 74 **In steps of** 1.

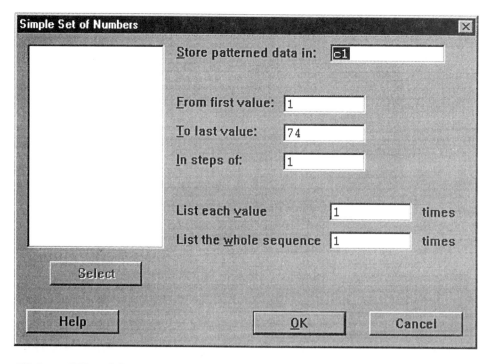

Click on **OK** and the numbers 1 to 74 should be in C1 of the Data Window.

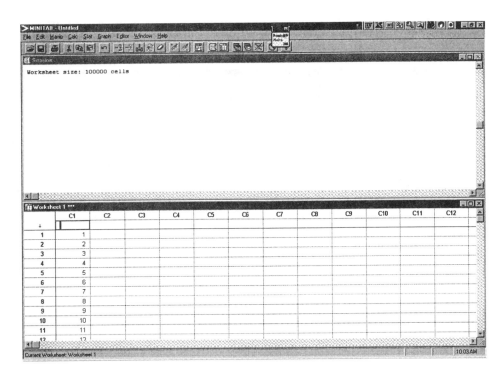

Next, you'd like to take a random sample of 8 accounts. Since you do not want repeats, you will be sampling without replacement. This is the default type of sampling in MINITAB, so you won't have to do anything special for this sample. Click on **Calc → Random Data → Sample from columns.** You need to **Sample 8 rows from column** C1 and **Store the sample in** C2.

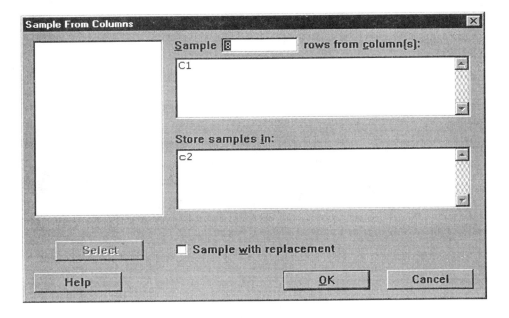

Click on **OK** and there should be a random sample of 8 account numbers in C2.

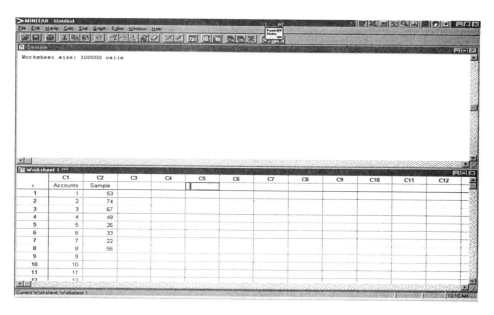

To order the sample list, click on **Manip → Sort**. You should **Sort column** C2 and **Store sorted column in** C3. You want to **Sort by column** C2.

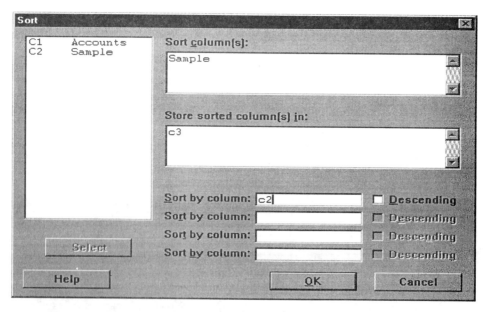

Click on **OK** and C3 should contain the sorted sample of 8 accounts.

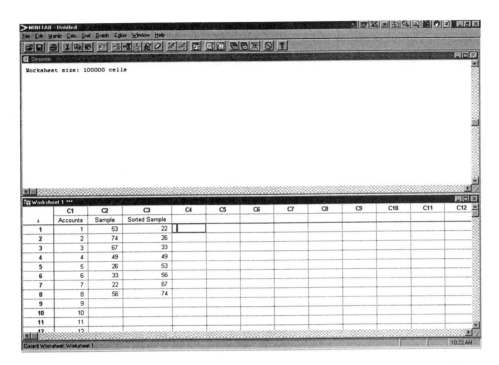

Since this is a *random* sample, each student will have different numbers in C2 and C3.

2. For this problem, use the same steps as above. Click on **Calc → Make Patterned Data → Simple Set of Numbers**. You should **Store patterned data in** C1. The numbers will begin **From the first value** 1 and go **To last value** 200 **In steps of** 1. Next, to randomly sample 20 batteries, click on **Calc → Random Data → Sample from columns**. You need to **Sample** 20 **rows from column** C1 and **Store the sample in** C2. Finally, to order the sample list, click on **Manip → Sort**. You should **Sort column** C2 and **Store sorted column in** C3. You want to **Sort by column** C2.

3. Click on **Calc → Make Patterned Data → Simple Set of Numbers**. You should **Store patterned data in** C1. The numbers will begin **From the first value** 0 and go **To last value** 9 **In steps of** 1. Next, to randomly sample 5 digits, click on **Calc → Random Data → Sample from columns**. You need to **Sample** 5 **rows from column** C1 and **Store the sample in** C2. Repeat this two more times and **store the sample in** C3 and then in C4. Now you have generated the three samples.

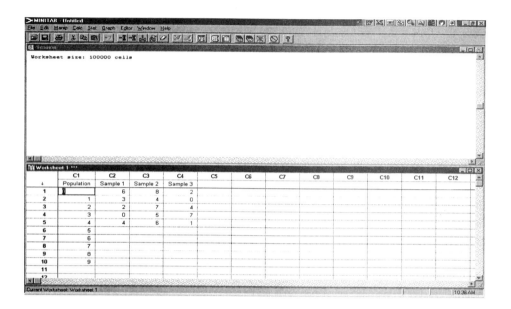

Now, to find the average of each of the four columns (C1, C2, C3, and C4), click on **Calc → Column Statistics**. The Statistic that you would like to calculate is the mean, so click on **Mean**. Enter C1 for the **Input variable** and click on **OK**.

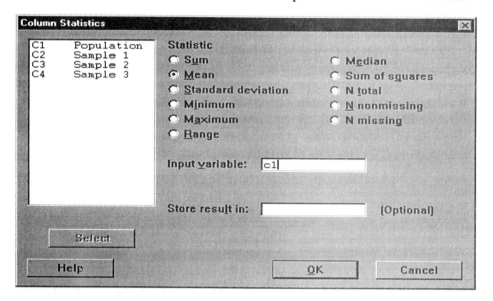

The population mean will be displayed in the Session Window. Repeat this for C2, C3, and C4.

Since this is random data, your output will look different.

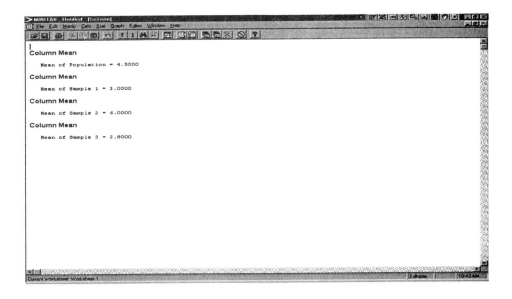

4. Repeat the steps in problem 3. Click on **Calc → Make Patterned Data → Simple Set of Numbers.** You should **Store patterned data in** C1. The numbers will begin **From the first value** 0 and go **To last value** 40 **In steps of** 1. Next, to randomly sample 7 digits, click on **Calc → Random Data → Sample from columns.** You need to **Sample** 7 **rows from column** C1 and **Store the sample in** C2. Repeat this two more times and **store the sample in** C3 and then in C4. Now you have generated the three samples. Now, to find the average of each of the four columns (C1, C2, C3, and C4), click on **Calc → Column Statistics.** The Statistic that you would like to calculate is the mean, so click on **Mean.** Enter C1 for the **Input variable** and click on **OK.** The population mean will be displayed in the Session Window. Repeat this for C2, C3, and C4.

5. To simulate rolling a 6-sided die, you want to sample with replacement from the integers 1 to 6. You could enter the numbers 1 to 6 into C1 and then sample with replacement. A better way to do this is to click on **Calc → Random Data → Integer.** You want to **Generate** 60 **rows of data** and **Store in column** C1. This represents the 60 rolls. Enter a **minimum value** of 1 and a **maximum value** of 6 to represent the six sides of the die.

14 Chapter 1 Introduction to Statistics

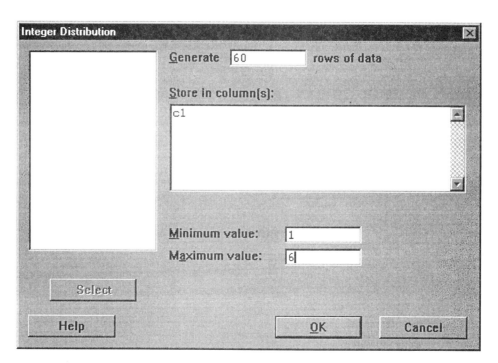

Click on **OK** and C1 should have the results of 60 rolls of a die.

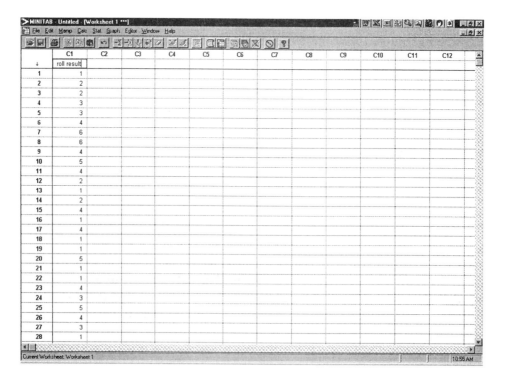

To count how many times each number was rolled, click on **Stat → Tables → Tally**. Enter C1 for the **Variable** and select **Counts** by clicking on it.

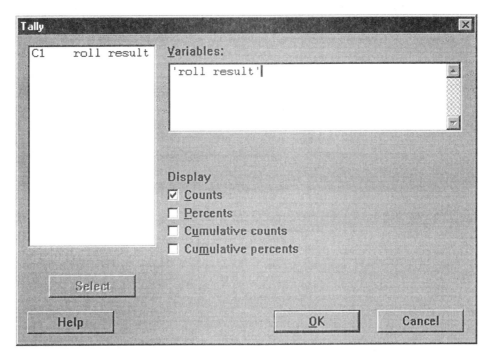

Click on **OK** and the totals will be displayed in the Session Window.

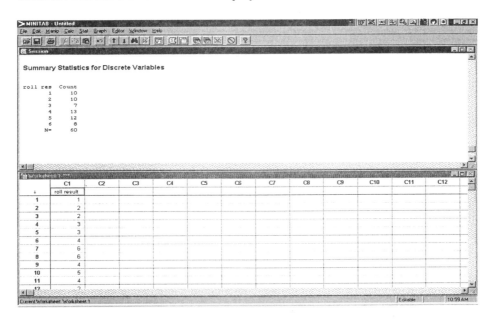

Recall that each student's results will look different because this is random data.

7. To simulate tossing a coin, you can repeat the steps in problem 5. Click on **Calc → Random Data → Integer.** You want to **Generate** 100 **rows of data** and **Store in column** C1. This represents the 100 tosses. Enter a **minimum value** of 0 and a **maximum value** of 1 to represent the two sides of the coin. Click on **OK** and C1 should have the results of 100 tosses of a coin. To count how many times each side of the coined was tossed, click on **Stat → Tables → Tally.** Enter C1 for the **Variable** and select **Counts** by clicking on it. Click on **OK** and the counts will be displayed in the Session Window. Recall that 0 represents heads and 1 represents tails.

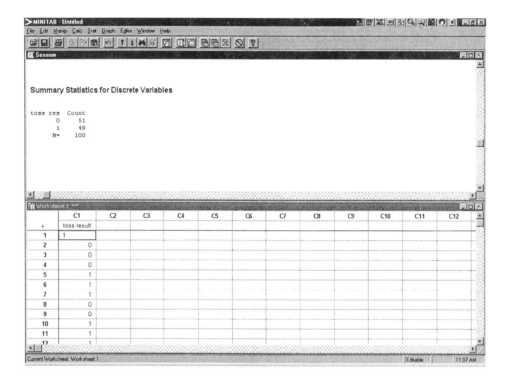

Descriptive Statistics

CHAPTER 2

Section 2.1

▶ **Example 7 (pg. 38):** Construct a histogram using the Internet data

To create this histogram, you must open the worksheet called **internet**. When you get into MINITAB, click on: **File → Open Worksheet**. On the screen that appears, **Look In** your 3½" Floppy (A:) to see the list of available folders. Double-click on the Folder that is named "**Minitab**" and then select the **Chapter 2** folder. Next click on the arrow to the right of the **Files of type** field and select **Minitab Portable (mtp)**. A list of data files should now appear. Use the right arrow to scroll through the list of files until you see **internet**. Click on this file, and then click on **Open**.

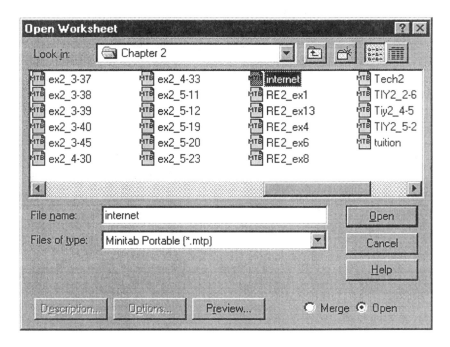

18 Chapter 2 Descriptive Statistics

You should now see Internet Subscriber data in Column 1.

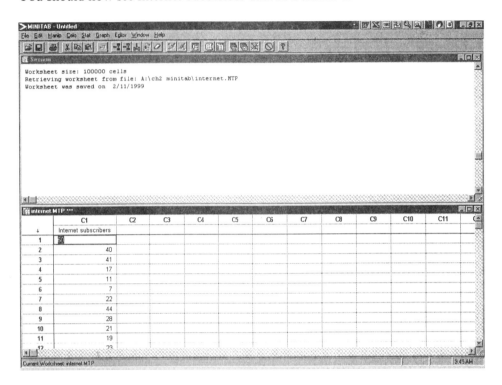

Now you are ready to make the histogram. Click on: **Graph → histogram**.
The following screen should appear.

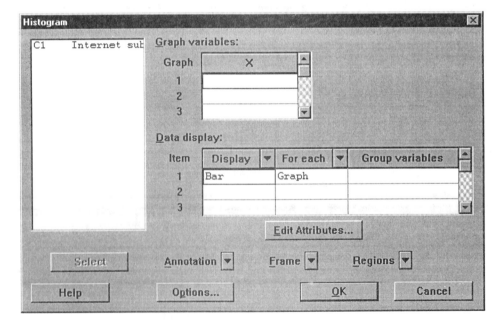

First, double click on C1 in the large box at the left of the screen. C1 should now be filled in as Graph 1.

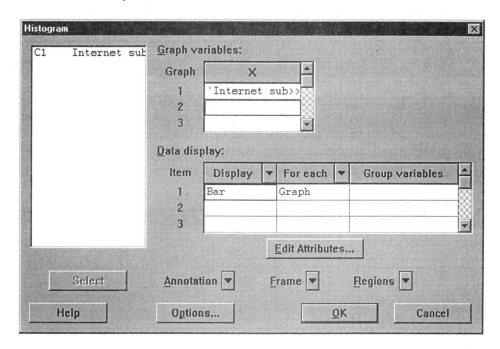

At this point, if you click on **OK**, MINITAB will draw a histogram using default settings. Your histogram will look like this.

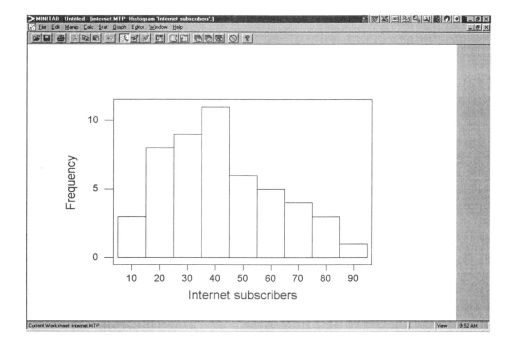

Notice the X-axis label is not "Minutes", and the numbering along the axis is not like the numbering in the textbook. We can instruct MINITAB to do this, however. Close the Graph Window by clicking on the "X" in the upper right corner of the graph. Go back to the main Histogram screen. (Click on: **Graph** → **histogram**.) To label the X-axis "Minutes", click on: **Frame** → **Axis**. You should see the following screen.

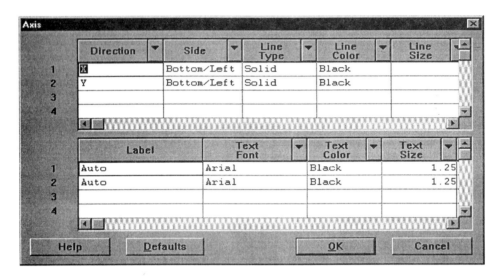

In the bottom half of the screen, beneath "Label" and beside "1", enter the label "Minutes" for the X-axis. To do this, click on the word "Auto" and delete it, then type in the new label.

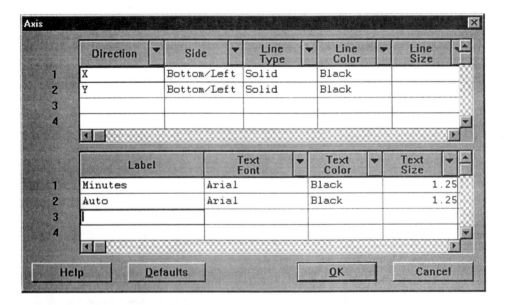

Click on **OK** to close the screen. You will now be back to the main Histogram screen. Next, fix the numbering along the X-axis. Notice the textbook uses the numbers: 12.5, 24.5, 36.5, etc. The numbers are centered below each rectangle of the histogram. To do this, simply click on **Options** and the following screen will appear.

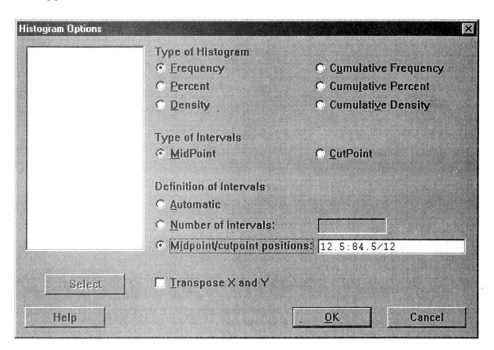

On the screen that appears, select **Frequency** as the **Type of Histogram** and **Midpoint** for **Type of Intervals**. Under **Definition of Intervals,** click on **Midpoint/cutpoint positions.** In the box to the right of it, type in: 12.5 : 84.5 / 12. This tells MINITAB that the first midpoint is 12.5 and the last midpoint is 84.5. The class width is 12, just as in the text. Click on **OK** to go back to the main Histogram screen, and then click on **OK** again to view your Histogram.

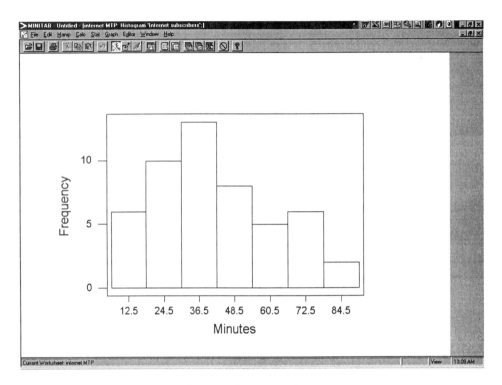

Now, your histogram is exactly as in the textbook.

To print the graph, click on **File → Print Graph.** Next, click **OK** and the graph should print.

Section 2.1   23

▶ **Exercise 17 (pg. 41)** Construct a frequency histogram using 6 Classes

Open the worksheet **ex2_1-17** which is found in the **Chapter 2** MINITAB folder. Select **Graph → Histogram.** Double click on C1, so that it appears beside Graph 1.

Next put a title on your graph, as any picture is incomplete without a title. Click on **Annotation → Title.** In line 1, type in "July Sales".

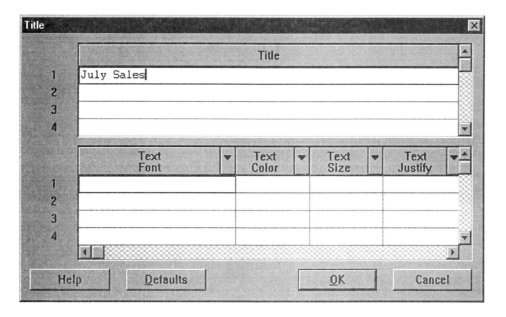

Click on **OK**, and you will be back to the main Histogram screen. Now change the axis label to "Dollars". To do this, click on **Frame** → **Axis**. Beneath **Label** in the bottom half of the screen, type "Dollars" into Line 1 for the X-axis.

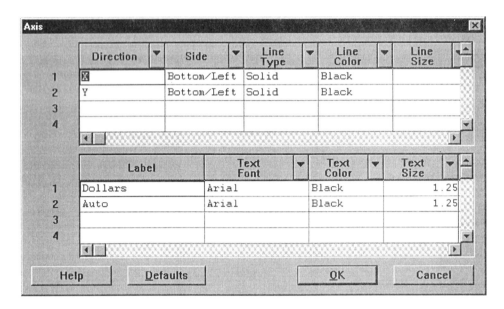

Next, decide on the numbering for the X-axis that is needed for the 6 classes. This will depend on the class limits you used in your frequency distribution. In order to use your class limits, you must tell MINITAB to use **Cutpoints.** To do this, click on **Options.** Select **Frequency** as the **Type of Histogram** and **CutPoint** as the **Type of Intervals.** Next tell MINITAB what the cutpoint positions are. One solution is to use cutpoints beginning at 1000 and going up to 7600 in steps of 1100. So fill in 1000:7600/1100 for **Midpoint/cutpoint positions**.

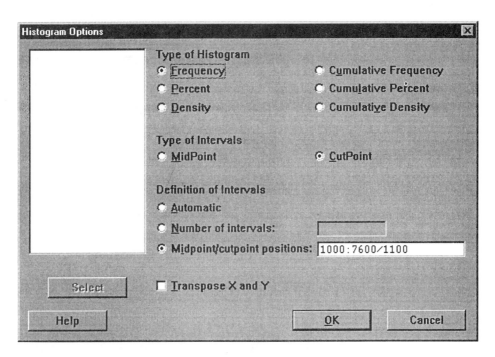

Click on **OK**. This will bring you back to the main Histogram screen. Click on **OK** again and you should see your histogram.

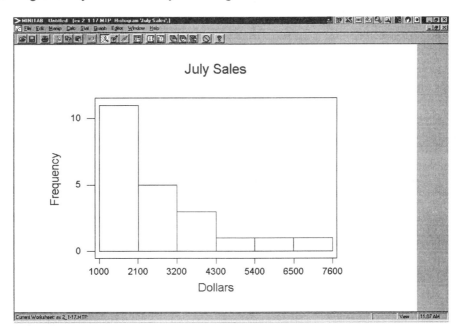

26 Chapter 2 Descriptive Statistics

> **Exercise 21 (pg.42)** Using the bowling scores, construct a relative frequency histogram with 5 classes

Open worksheet **ex2_121** which is found in the **Chapter 2** MINITAB folder. Click on **Graph → histogram**. Double-click on C1 to select it for the graph. Add a title to the histogram by clicking on **Annotation → title**. In Line 1, enter "Bowling Scores". Next, enter a label for the X-axis by clicking on **Frame → Axis**. Beneath **Label**, on Line 1, enter "Points scored". Now click on **Options**. Since this is a relative frequency histogram, the **Type of Histogram** should be **Percent.** Next the cutpoints for the classes must be chosen. One solution is to use 145 to 270 in steps of 25 as cutpoints. So enter 145:270/25 in the appropriate place.

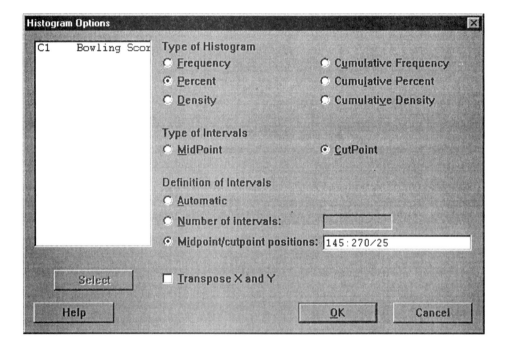

Click on **OK** twice, and the histogram should appear.

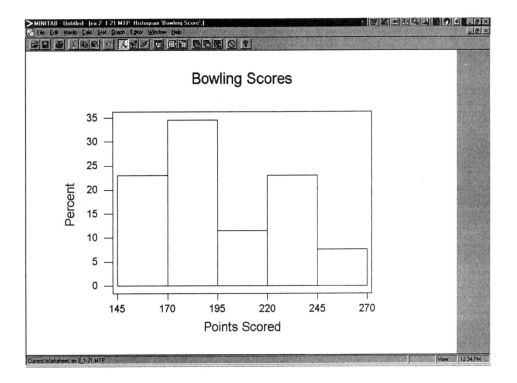

▶ **Exercise 31a (pg. 43)** Construct a relative frequency histogram for the ATM data using 8 classes

Open worksheet **ex2_1-31** which is found in the **Chapter 2** MINITAB folder. Click on **Graph → Histogram.** Double-click on C1 to select it for the graph. Add a title to the histogram by clicking on **Annotation → title**. In Line 1, enter "Daily Withdrawals". Next, enter a label for the X-axis by clicking on **Frame → Axis.** Beneath **Label**, on Line 1, enter "Hundreds of Dollars". Click on **Options** and select **Percent** as the **Type of Histogram**. Next the cutpoints for the classes must be chosen. One possible solution is to use 60 to 108 in steps of 6 as cutpoints. So enter 60:108/6 in the appropriate place. Click on **OK** twice and the histogram should appear.

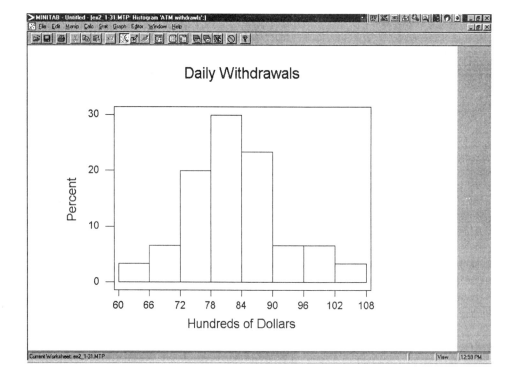

Section 2.2

> **Example 2 (pg. 45)** Constructing a Stem-and-Leaf Plot

Open the file **AL_RBIs** which is found in the **Chapter 2** MINITAB folder. This worksheet contains data on runs batted in (RBIs) for baseball's American League for the last 50 years. The data should appear in C1 of your worksheet.

To construct a Stem-and-leaf plot, click on **Graph → Stem and Leaf**.
On the screen that appears, select C1 as your **Variable** by doubling clicking on C1. Click on **OK**.

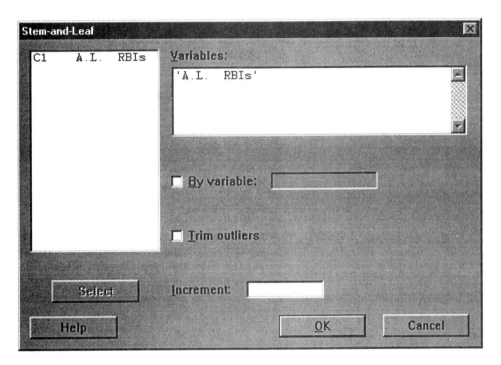

The stem and leaf plot will be displayed in the Session Window.

```
Character Stem-and-Leaf Display

Stem-and-leaf of A.L.   RB   N = 50
Leaf Unit = 1.0

     1     7  8
     1     8
     1     8
     1     9
     1     9
     1    10
     6    10  58999
    11    11  22234
    19    11  67888999
    (8)   12  11222344
    23    12  6666699
    16    13  2334
    12    13  99
    10    14  0024
     6    14  5578
     2    15
     2    15  59
```

In this MINITAB display, the first column on the left is a counter. This column counts the number of data points starting from the smallest value (at the top of the plot) down to the median. It also counts from the largest data value (at the bottom of the plot) up to the median. Notice that there is only one data point in the first row of the stem and leaf. There are no data points in rows 2 through 6, so the counter on the left remains at "1". Row 7 has 5 data points so the counter increases to "6". The row that contains the median has the number "8" in parentheses. This number counts the number of data points that are in the row that contains the median.

The second column in the display is the **Stem**. In this example, the Stem values range from 7 to 15. Notice that this display contains two rows for each of the values from 8 through 15. These are called **split-stems**. For each stem value, the first row contains all data points with leaf values from 0 to 4 and the second row contains all data points with leaf values from 5 to 9. Notice that MINITAB constructs an *ordered* stem and leaf.

The leaf values are shown to the right of the stem. The leaf values may be the actual data points or they may be the rounded data points. To find the actual values of the data points in the display, use the "Leaf Unit=" statement at the top of the display. The "Leaf Unit" gives you the place value of the leaves. In this stem and leaf, the first data point has a stem value of 7 and a leaf value of 8. Since the "Leaf Unit=1.0", the 8 is the "ones" place and the 7 is in the "tens" places, thus the data point is 78.

Section 2.2   31

▶ Example 3 (pg. 46)   Constructing a Dotplot

Open the file **AL_RBIs** which is found in the **Chapter 2** MINITAB folder. To construct a dotplot, click on **Graph → Dotplot**. In the screen that appears, select C1 for **Variables**. Since there is only one column of data, be sure that **No grouping** is selected. Enter an appropriate **Title** and click on **OK**.

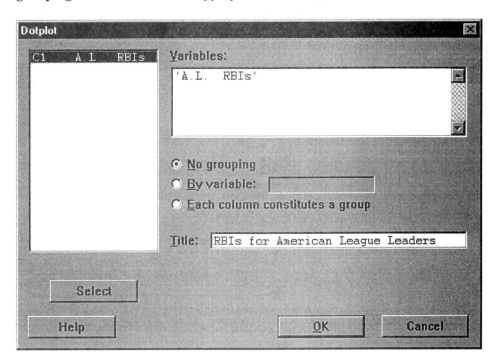

The following dotplot should appear.

32 Chapter 2 Descriptive Statistics

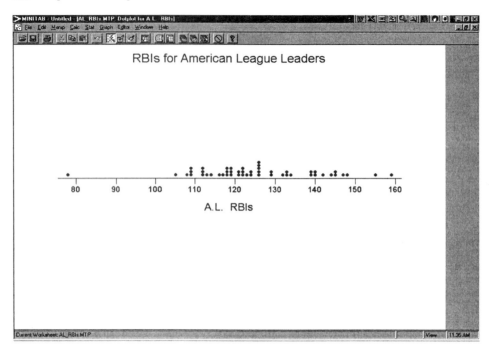

Each dot in the above plot represents the RBIs for an individual batter. For example, one batter had 78 RBIs, and 3 batters had 109 RBIs (the 3 dots above 109).

Section 2.2 33

Example 4 (pg. 47) Constructing a Pie Chart

In this example, you must enter the data into the Data Window. Begin with a clean worksheet. From the table in the left margin of page 47, enter the Transportation types into C1 and the number of Passengers (in millions) into C2. Label each column appropriately as shown below.

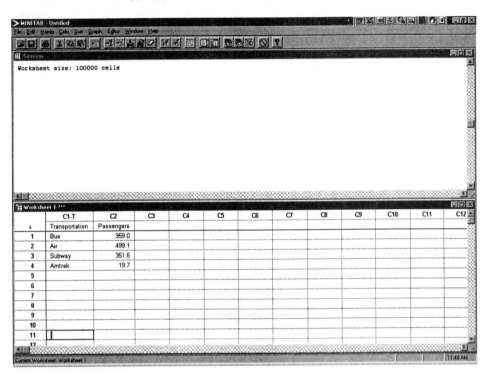

To construct the pie chart, click on **Graph → Pie Chart**. In the screen that appears, select **Chart Table**. Fill in the screen as follows: **Categories In** C1 and **Frequencies In** C2. Click on **Title** and enter an appropriate title.

34 Chapter 2 Descriptive Statistics

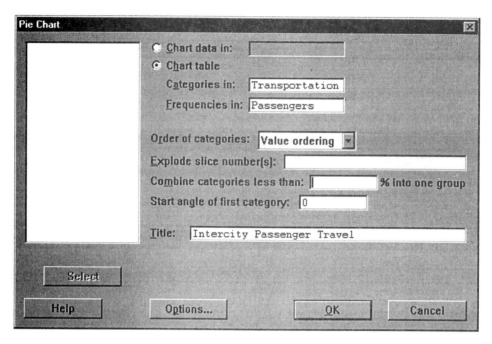

Click on **OK** to view the pie chart.

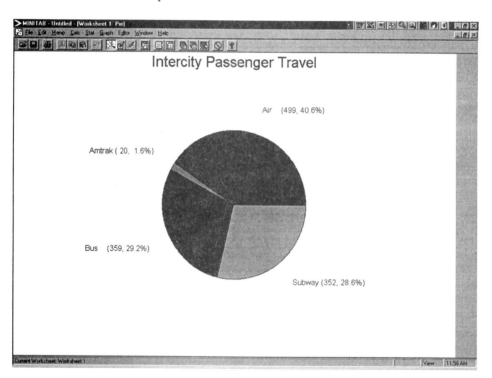

Example 5 (pg. 48) Constructing Pareto Charts

Enter the Inventory Shrinkage data (found in the paragraph for Example 5) into C1 and C2. Do not include the Total amount of $40.9 million.

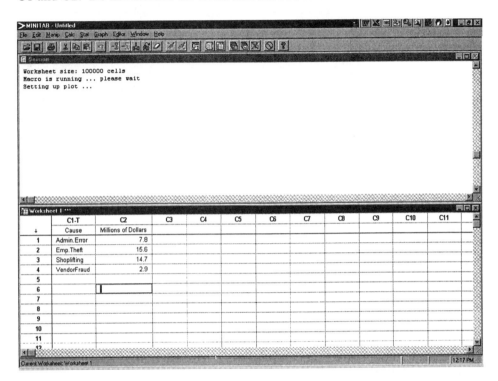

To make the Pareto chart, click on **Graph → Chart**. Select C2 for the **Y** variable and C1 for the **X** variable. (Leave **Function** blank.)

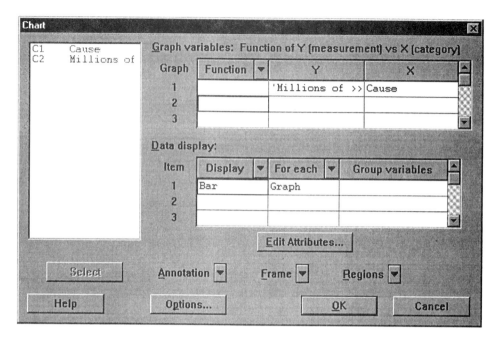

Click on **Annotation** → **Title** and enter an appropriate title for the chart. Click on **Frame** → **Axis** and enter "Millions of Dollars" for the Y-axis label and "Cause" for the X-axis label. Next, set the scale for the Y-axis to go from 0 to 16 in steps of 2 as shown in the text. Click on **Frame** → **Tick**. For **Y Positions**, enter 0:16/2. Click on **OK**.

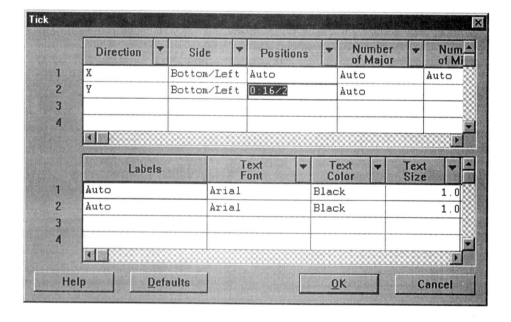

Notice that the bars are connected in a Pareto Chart. MINITAB does not do this automatically, so you will have to change the default. Click on **Edit Attributes**. Use the right arrow to page over to **Bar Width**. Replace "Auto" with "1". Click on **OK**.

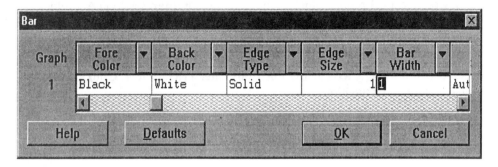

To position the bars in decreasing order, click on **Options** and select **Decreasing Y**. Click on **OK** twice and view the chart.

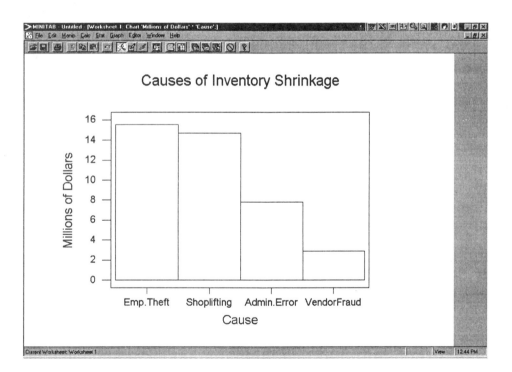

38 Chapter 2 Descriptive Statistics

Example 7 (pg. 50) Construct a Time Series Chart of cellular telephone subscribers

Open worksheet **cellphone** which is found in the **Chapter 2** MINITAB folder. Click on **Graph→ Plot**. Select C2 as the **Y variable** and C1 as the **X variable**. Click on **Annotation → Title** and enter an appropriate title for the plot. Next, click on **Frame → Tick**. You would like to see a tick mark on the x-axis for each year, so for the **X axis** below **Positions** enter 1987:1996/1. Click on **OK** twice, and the view the plot.

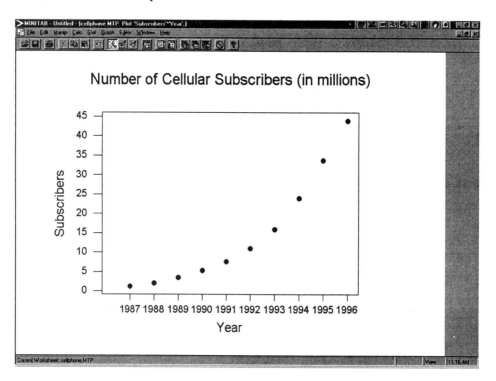

Exercise 17 (pg. 52) Construct a stem and leaf of the amount of water consumed by 24 elephants in one day

Open worksheet **ex2_2-17** which is found in the **Chapter 2** MINITAB folder. Click on **Graph→ Stem-and-Leaf.** Select C1 for the **Variable.** Click on **OK** and the stem and leaf plot should be in the Session Window.

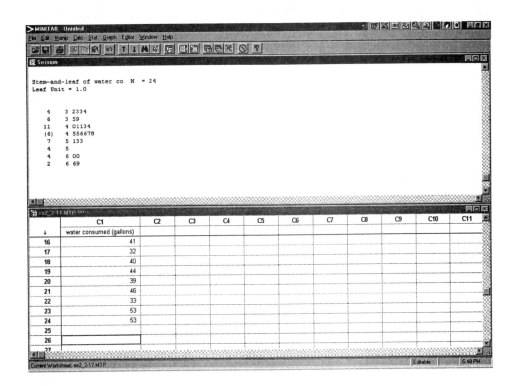

Exercise 21 (pg. 53) Construct a dotplot of the lifespan (in days) of houseflies

Open worksheet **ex2_2-21** which is found in the **Chapter 2** MINITAB folder. Click on **Graph→ Dotplot.** Select C1 for the **Variable** and enter an appropriate **Title.** Click on **OK** to view the dotplot.

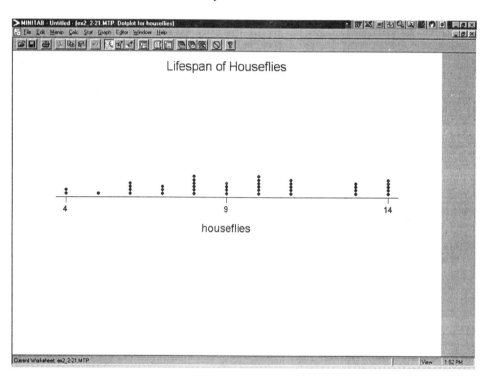

▶ Exercise 23 (pg. 53) Construct a Pie Chart of the data.

This data must be entered into the Data Window. Enter the categories into C1 and the Expenditures into C2. Click on **Graph → Pie Chart.** Select **Chart Table**. Enter **Categories In** C1 and **Frequencies In** C2. Enter an appropriate **Title** and click on **OK**.

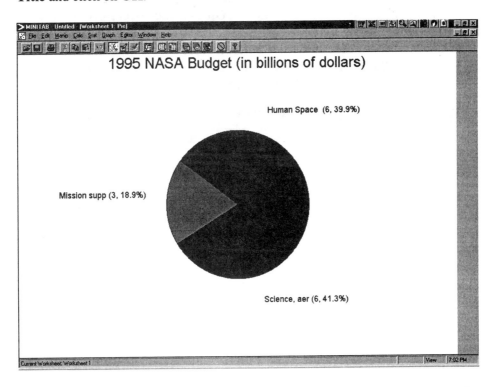

Exercise 25 (pg. 53) Construct a Pareto Chart to display the data

Enter the data into the Data Window. Enter the cities into C1 and the ultraviolet index for each city into C2. Click on **Graph → Chart**. Select C2 as the **Y** variable and C1 as the **X** variable. Click on **Options** and select **Decreasing Y**. Click on **OK**. Click on **Edit Attributes** and enter a "1" for **Bar Width**. Click on **OK**. Click on **Annotation → Title**. Enter an appropriate title. Click on **Frame → Axis**. Enter "Index" for the **Y-axis Label** and "City" for the **X-axis Label**. Click on **OK** twice to view the Pareto Chart.

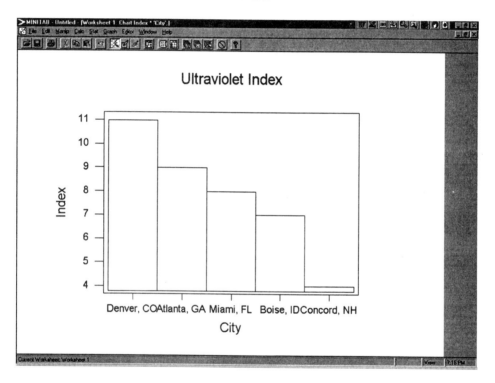

Section 2.2 43

▶ **Exercise 27 (pg. 54)** Construct a Scatterplot of the data.

Enter the Number of Students per Teacher into C1 and the Average Teacher's Salary into C2. Click on **Graph → Plot.** Select C2 for the **Y variable** and C1 for the **X variable.** Click on **Frame → Axis.** Enter 25:55/5 for the **Y Positions.** Click on **Annotation → Title** and enter an appropriate title. Click on **OK** twice to view the scatterplot.

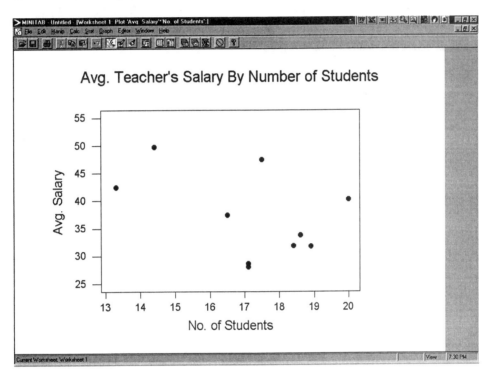

44 Chapter 2 Descriptive Statistics

Exercise 29 (pg. 54) Construct a Time Series Plot of the price of milk.

Enter the data into the Data Window. Enter the Years into C1 and the Price of Milk into C2. Click on **Graph → Plot.** Select C2 as the **Y variable** and C1 as the **X variable.** Click on **Annotation → Title.** Enter an appropriate title. Click on **OK** twice to view the plot.

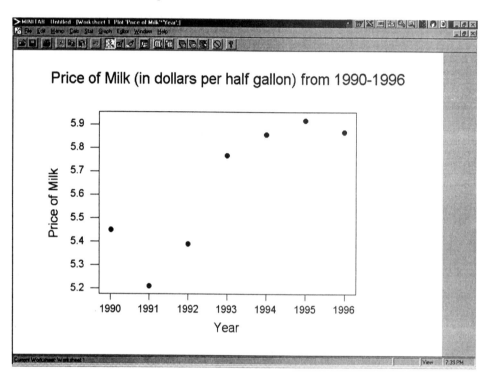

Section 2.3

▶ **Example 6 (pg. 58)** Find the mean and standard deviation of the age of students

Finding the mean and standard deviation of a dataset is very easy using MINITAB. Open the worksheet **ages** which is found in the **Chapter 2 MINITAB folder.** Click on **Stat → Basic Statistics → Display Descriptive Statistics.** You should see the input screen below.

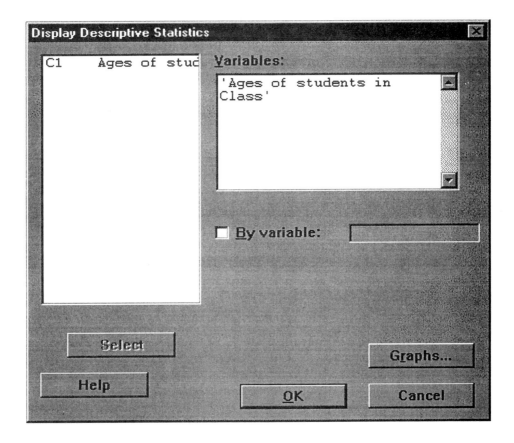

Double click on C1 to select the age data that is entered in C1. Click on **OK** and the descriptive statistics should appear in the Session Window.

46 Chapter 2 Descriptive Statistics

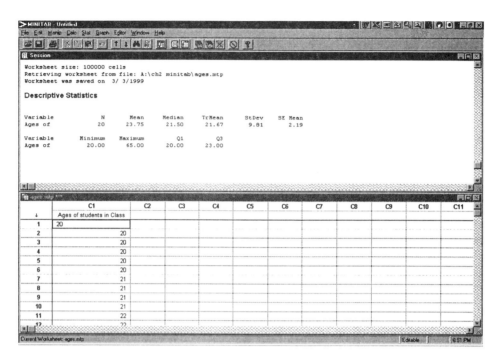

Notice that MINITAB displays several descriptive statistics: sample size, mean, median, trimmed mean, standard deviation, standard error of the mean, minimum value, maximum value, and the first and third quartiles.

The mode is NOT produced by the above procedure, however, it is quite simple to have MINITAB tally up the data values for you, and then you can select the one with the highest count. Click on **Stat** → **Tables** → **Tally**. On the input screen, double-click on C1 to select it. Also, click on **Counts** to have MINITAB count up the frequencies for you.

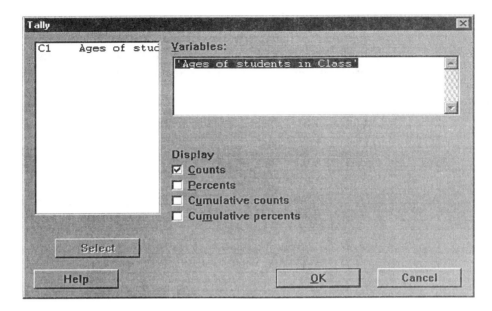

When you click on **OK**, a frequency table will appear in the Session Window.

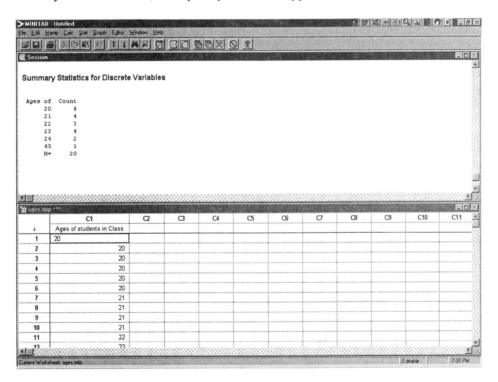

Notice that Age 20 has a count of 6. This means that 6 people in the data were age 20. Since this is the highest count, 20 is the mode.

To print the Session Window with both the descriptive statistics and the frequency table in it, click anywhere up in the Session Window to be sure that it is the active window. Next click on **File → Print Session Window**.

Exercise 17 (pg. 63) Find the mean, median, mode for points per game scored by each NBA team

Open worksheet **ex2_3-17** which is found in the **Chapter 2** MINITAB folder. Click on **Stat → Basic Statistics → Display Descriptive Statistics**. Double-click on C1 to select it. Click on **OK** and the results should be in the Session Window. Next, make the frequency table. Click on **Stat → Tables → Tally**. Double-click on C1, then click on **OK**. Now both of the displays will be in the Session Window.

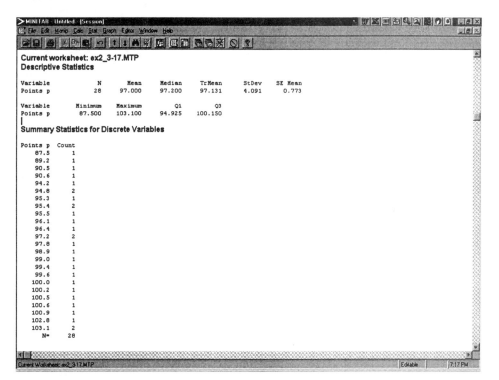

Exercise 39 (pg. 66) Construct a frequency histogram of the heights using 5 classes

Open worksheet **ex2_3-39** which is found in the **Chapter 2** MINITAB folder. Click on **Graph → Histogram**. One possible solution is to use cutpoints 62 to 77 in steps of 3. Click on **Options** and select **Frequency** as the **Type of Graph**, select **cutpoints**, and then enter 62:77/3. Click on **OK** twice and the following histogram should be displayed.

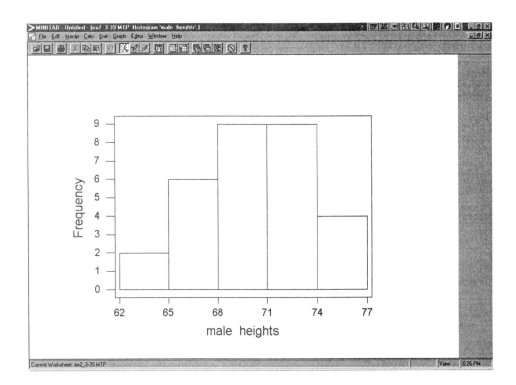

▶ Exercise 45 (pg. 67) Find the mean, median, and construct a stem and leaf plot of the data

Open worksheet **ex2_3-45** which is found in the **Chapter 2** MINITAB folder. Click on **Stat → Basis Statistics → Display Descriptive Statistics.** Select C1 for the **Variable** and click on **OK.** The descriptive statistics should be in the Session Window. Next, click on **Graph → Stem-and-Leaf.** Select C1 for the **Variable** and click on **OK.** The stem and leaf plot should also be in the Session Window as shown in the next picture.

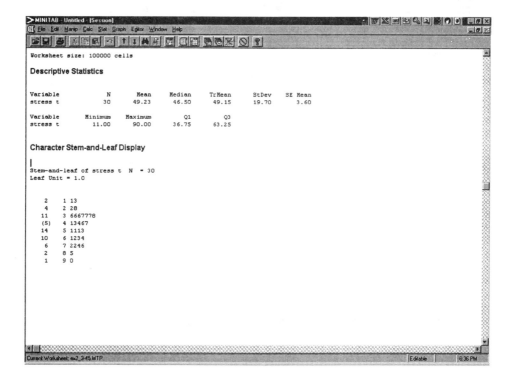

Section 2.4

▶ **Example 5 (pg. 72)** Calculate the mean and standard deviation for the office rental rates in Atlanta

Open worksheet **atlanta** which is found in the **Chapter 2** MINITAB folder. Click on **Stat → Basis Statistics → Display Descriptive Statistics**. Select C1 for the **Variable** and click on **OK**. The descriptive statistics should be in the Session Window.

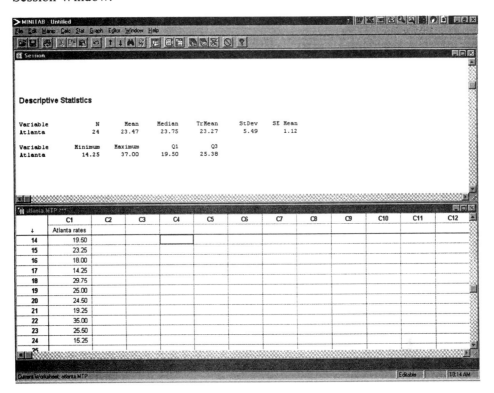

Section 2.4 53

▶ Exercise 11 (pg. 79) Find the range, mean, variance, standard deviation of the dataset.

Enter the data into C1 of the Data Window. Click on **Stat → Basis Statistics → Display Descriptive Statistics**. Select C1 for the **Variable** and click on **OK**. The descriptive statistics should be in the Session Window.

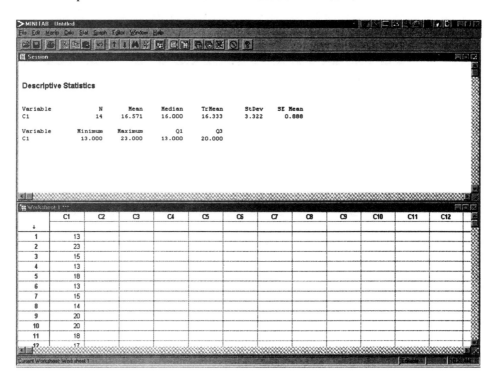

Exercise 17 (pg. 79) Find the mean, range, standard deviation and variance for each city

Enter the data into C1 and C2 of the Data Window. Be sure to label the columns. Click on **Stat → Basis Statistics → Display Descriptive Statistics.** Select both C1 and C1 for the **Variable** and click on **OK.** The descriptive statistics should be in the Session Window.

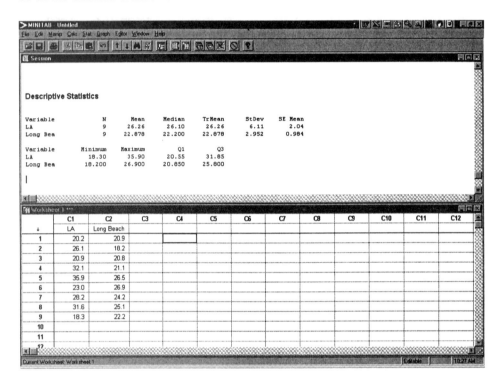

Section 2.5

▶ Example 2 (pg. 86) Find the first, second, and third quartiles of the tuition data.

Open worksheet **tuition** which is found in the **Chapter 2** MINITAB folder. Click on **Stat → Basis Statistics → Display Descriptive Statistics.** Select C1 for the **Variable** and click on **OK.** The descriptive statistics should be in the Session Window. (Note that the median is the second quartile.)

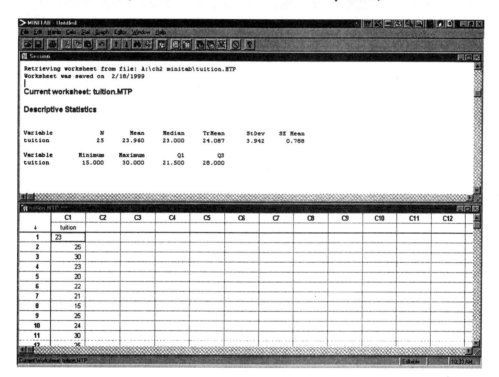

56 Chapter 2 Descriptive Statistics

▸ **Example 4 (pg. 88)** Construct a box-and-whisker plot using the test scores given in Example 1.

Enter the data found on page 85 of the text into C1 of the Data Window. Click on **Graph → Boxplot**. Select C1 for the **Y variable**. Next, since by default MINITAB plots vertically, click on **Options** and select **Transpose X and Y**. This will turn the plot horizontal, as in the textbook. Click on **OK** twice to view the box-and-whisker plot.

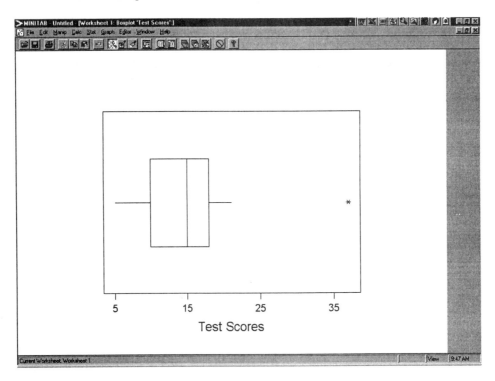

If you want to change the tick marks below the box-and-whisker plot so that the numbering will include more tick marks, from the main boxplot screen, (Graph → Boxplot) click on **Frame → Tick**. Below **Positions** and beside **Y**, enter tick positions 5:35/5. This tells MINITAB to place tick marks beginning at 5 and ending at 35, in steps of 5. With more tick marks, it is easier to read where the quartiles and median actually are.

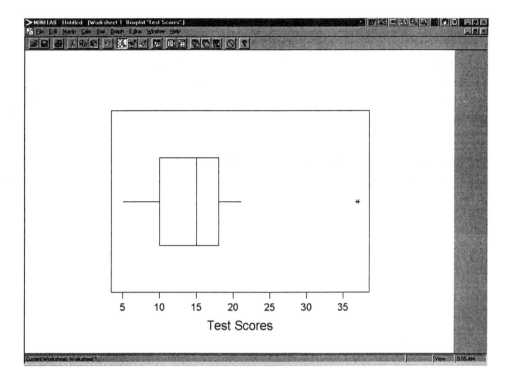

58 Chapter 2 Descriptive Statistics

▶ **Exercise 11 (pg. 90)** Draw a box-and-whisker plot of the data

Enter the data into C1 of the Data Window. Give C1 an appropriate label, such as "Exercise 11". Click on **Graph → Boxplot**. Select C1 for the **Y variable.** Next, since by default MINITAB plots vertically, click on **Options** and select **Transpose X and Y.** This will turn the plot horizontal, as in the textbook. Click on **OK** twice to view the box-and-whisker plot.

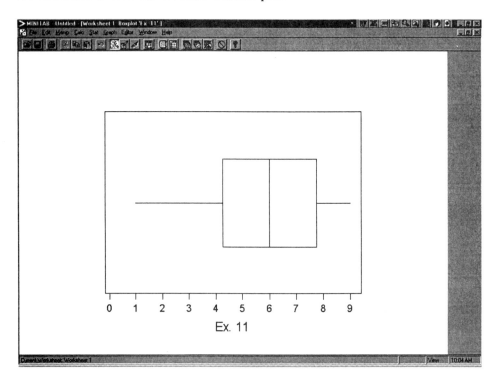

▶ **Exercise 19 (pg. 91)** Draw a box-and-whisker plot of hours of TV watched per day

Open worksheet **ex2_5-19** which is found in the **Chapter 2** MINITAB folder. Click on **Graph→ Boxplot.** Select C1 for the **Y variable.** Because MINITAB's default setting plots vertically, you will need to click on **Options** and select **Transpose X and Y.** This will turn the plot horizontal, as in the textbook. Click on **OK** twice to view the box-and-whisker plot.

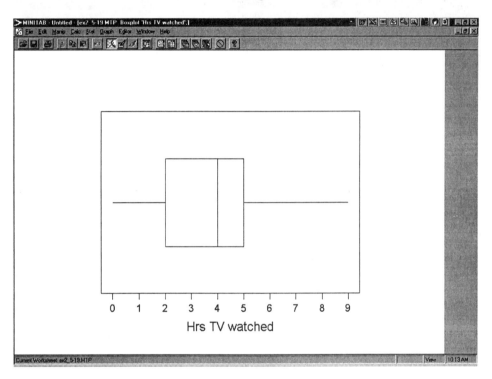

▶ Technology Lab (pg. 93) Use descriptive statistics and a histogram to describe the milk data

Open worksheet **Tech2** which is found in the **Chapter 2** MINITAB folder. Click on **Stat → Basic Statistics → Display Descriptive Statistics.** Click on **OK** and the descriptive statistics will be displayed in your Session Window.

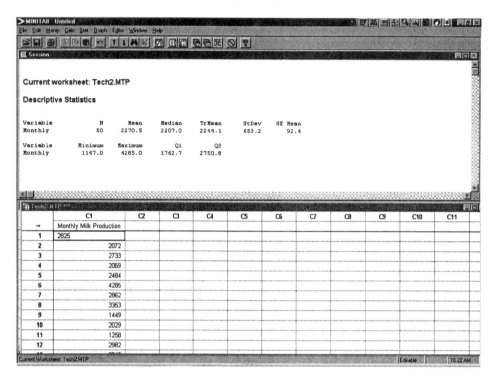

Next construct the histogram. Click on **Graph→ Histogram.** Double click on C1 to enter it for the **X variable.** Click on **Annotation → Title** and enter an appropriate title for the histogram. Click on **OK.** Next click on **Options.** Select **Frequency** for the **Type of Histogram, Cutpoint** for the **Type of Intervals.** Using the information from the descriptive statistics, you can see that the minimum value is 1147 and the maximum is 4285. Since you want a class width of 100, use positions beginning at 1100 and going up to 4300 in steps of 500. Thus, enter the **Midpoint/cutpoint positions** as **1100:4300/500.** Click on **OK.** Next, you should place the tick marks. Click on **Frame→ Tick.** In the first row, beneath **Positions,** enter 1100:4300/500. Click on **OK** twice to view the histogram.

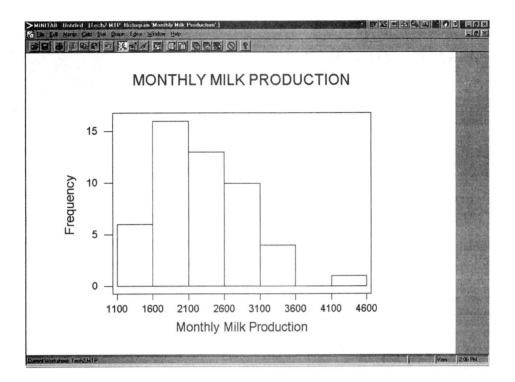

Probability

Section 3.1

> **Law of Large Numbers (pg. 108)** Coin Simulation

You can use MINITAB to simulate repeatedly tossing a fair coin and then calculate the empirical probability of tossing a head. This empirical probability will more closely approximate the theoretical probability as the number of tosses gets large. To do this simulation, generate 1000 "tosses" of a fair coin. Let "0" represent a head, and "1" represent a tail. Click on **Calc → Random Data → Bernoulli**. **Generate** 1000 **rows of data** and **Store in** C1. Use .5 for the **Probability of Success.** When you click on **OK**, you should see C1 filled with 1's and 0's. To count up the number of "0"s, click on **Stat → Tables → Tally**. Select C1, and choose both **counts and percents** by clicking on the box to the left of each one. When you click on **OK**, the summary statistics will appear in the Session Window. In the following example, notice that there were 511 heads and 489 tails. Thus, the empirical of tossing a head is .511 (51.1%). This is a very good approximation of the theoretical probability of .5.

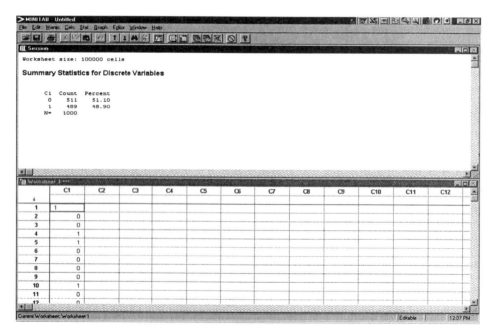

Section 3.2

> **Exercise 25 (pg. 123)** Birthday Problem

Simulate the "Birthday Problem" using MINITAB. To do this simulation, the days of the year will be represented by the numbers 1 to 365. Click on **Calc → Random Data → Integer.** Generate 24 **rows of data** (representing the 24 people in the room), **Store In** C1, **Minimum value** is 1 (representing Jan.1st) and **Maximum value** is 365 (representing Dec. 31st). The question is: Are there at least two people in the room with the same birthday? To answer this question, summarize the data. To do this, click on **Stat → Tables → Tally.** On the screen that appears, select C1 and click on **Counts.** Next click on **OK**, and in the Session Window, you will see a summary table of C1. This table lists each of the different birthdays that occurred in this simulation, as well as a count of the number of people who had that birthday. Notice that most counts are 1's. If you see a count of "2" or more, then you have at least two people in the room with the same birthday.

Repeat the simulation 9 more times. How many of the 10 columns had at least two people with the same birthday? The empirical probability that at least two people in a room of 24 people will share a birthday can be calculated as follows: (# of columns with matching pair / 10).

▶ **Technology Lab (pg. 143)** Composing Mozart Variations

3. Click on **Calc** → **Random Data** → **Integer**. **Generate** 1 row of data, **Store In** C1, **Minimum value** is 1 and **Maximum value** is 11. For Part B, **Generate** 100 **rows of data.** To tally the results, click on **Stat** → **Tables** → **Tally.** Select both **Counts** and **Percents.** The results will appear in the Session Window. Compare the percents to the theoretical probabilities you found in Part A.

5. Click on **Calc** → **Random Data** → **Integer**. **Generate** 2 **rows of data, Store In** C1, **Minimum value** is 1 and **Maximum value** is 6. Add up to two numbers and subtract 1 to obtain the total. For Part B, **Generate** 100 **rows of data, Store In** C1-C2, **Minimum value** is 1 and **Maximum value** is 6. The total will be calculated for each row by adding the two numbers from C1 and C2 and then subtracting 1. To do this, click on **Calc** → **Calculator**. **Store result in variable** C3 after calculating the following **Expression:** C1 + C2 - 1. Click on **OK**, and the totals should be in C3. To tally the results, click on **Stat** → **Tables** → **Tally.** Select both **Counts** and **Percents.** The results will appear in the Session Window. Compare the percents to the theoretical probabilities you found in Part A.

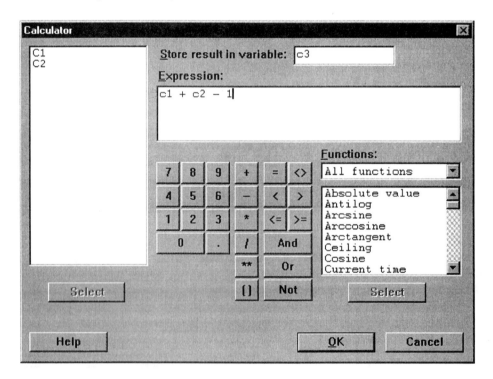

7. To choose a minuet, Mozart suggested that the player toss a pair of dice 16 times. For the 8th and 16th bars, choose Option 1 if the dice total is odd, and Option 2 if the dice total is even. For each of the other 14 bars, subtract 1 from the dice total. To do this in MINITAB, first simulate rolling the dice 16 times. Click on **Calc → Random Data → Integer. Generate** 16 **rows of data, Store In** C1-C2, **Minimum value** is 1 and **Maximum value** is 6. The total will be calculated for each row by adding the two numbers from C1 and C2 and then subtracting 1. To do this, click on **Calc → Calculator. Store result in variable** C3 after calculating the following **Expression:** C1+C2-1. Click on **OK**, and the totals should be in C3.

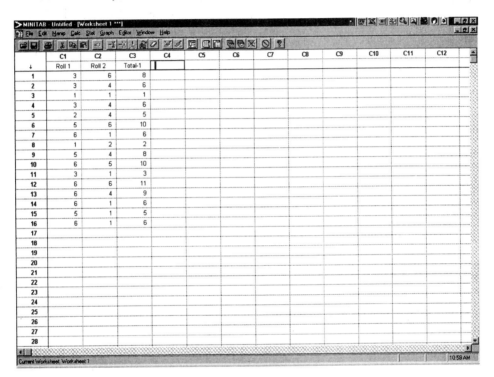

The numbers in C3 will be the minuet, except for the 8th and 16th bars. To find these, add C1 + C2 for rows 8 and 16. If the total is odd, choose Option 1 and if the total is even, choose Option 2. For example, the total in row 8 is 3 and Option 1 should be chosen. The total in row 16 is 7, and Option 1 should be chosen again. Thus, the minuet for this simulation is:

8	6	1	6	5	10	6	1
8	10	3	11	9	6	5	1

Notice the 8th and 16th bars are both 1.

Discrete Probability Distributions

Section 4.2

Example 5 (pg. 169) Find the probability that 65 out of 100 households own a gas grill

In this example, 58% of American households own a gas grill and a random sample of 100 American households is selected. Thus n = 100 and p = .58. Click on **Calc → Probability Distributions → Binomial.** To find the probability that exactly 65 of the 100 households own a gas grill, select **Probability**. This tells MINITAB what type of calculation you want to do. The **Number of Trials** is 100 and the **Probability of Success** is .58. To find the probability of 65, enter 65 beside **Input Constant**. Leave all other fields blank. Click on **OK**.

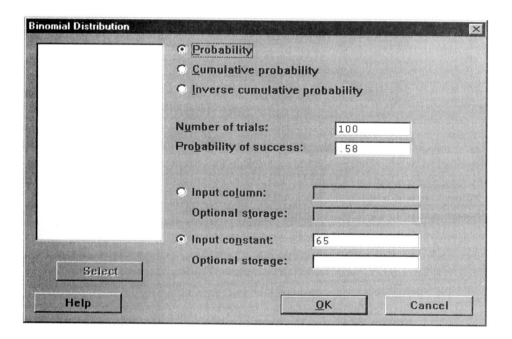

The probability that 65 of the 100 households sampled own a gas grill will be displayed in the Session Window. Notice that the probability is .0299.

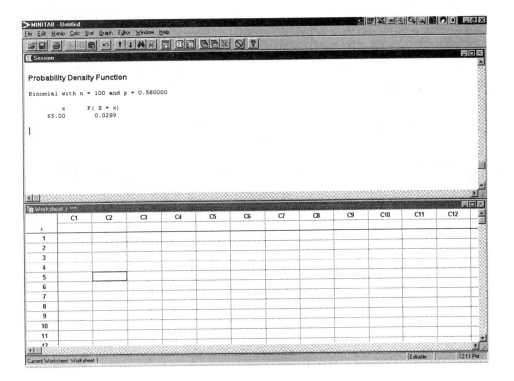

68 Chapter 4 Discrete Probability Distributions

▶ **Example 7 (pg. 171)** Constructing and Graphing a Binomial Distribution

In order to graph the binomial distribution, you must first create the distribution and save it in the Data Window. In C1, type in the values of X. Since n=6, the values of X are 0, 1, 2, 3, 4, 5, and 6. Next, use MINITAB to generate the binomial probabilities for n=6 and p=.65. Click on **Calc → Probability Distributions → Binomial**. Select **Probability**. The **Number of Trials** is 6 and the **Probability of Success** is .65. Now, tell MINITAB that the X values are in C1 and that you want the probabilities stored in C2. Enter C1 as the **Input Column** and enter C2 for **Optional Storage**.

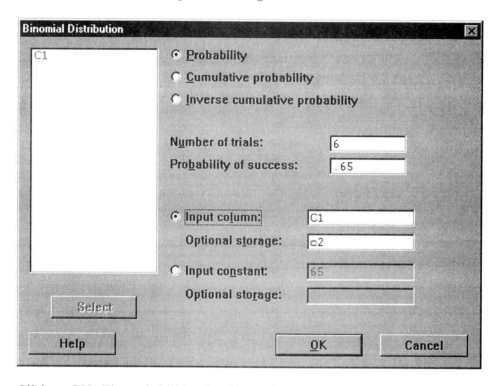

Click on **OK**. The probabilities should now be in C2. Label C1 as "X" and C2 as "P(X)". This will be helpful when you graph the distribution.

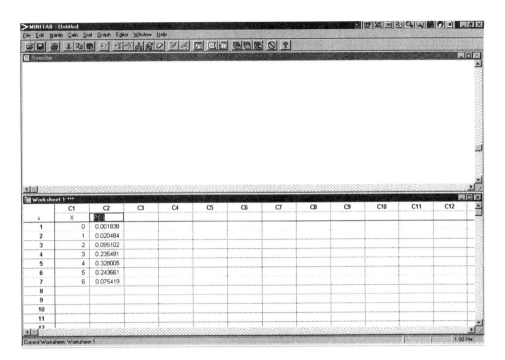

To create the graph, click on **Graph → Chart.** Enter C2 for the **Y** variable and C1 for the **X** variable. Leave **Function** blank. Click on **Annotation → Title** and enter an appropriate title to identify the binomial distribution. Click on **OK** twice to display the graph.

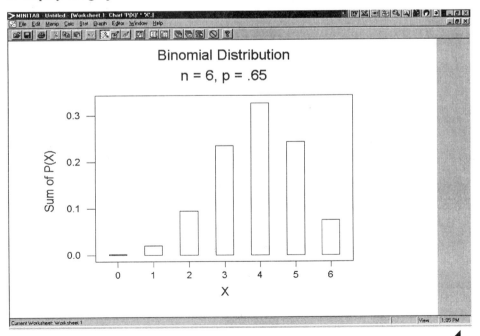

70 Chapter 4 Discrete Probability Distributions

Section 4.3

▶ **Example 3 (pg. 180)** Finding Poisson Probabilities

Since there is an average of 3.6 rabbits per acre living in a field, $\mu = 3.6$ for this Poisson example. To find the probability that 2 rabbits are found on any given acre of the field, click on **Calc → Probability Distributions → Poisson**. Since you want a simple probability, select **Probability** and enter 3.6 for the **Mean**. To find the probability that X=2, enter 2 for the **Input constant**.

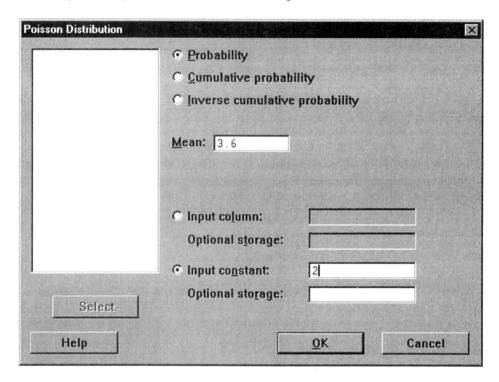

Click on **OK** and the probability will be in the Session Window.

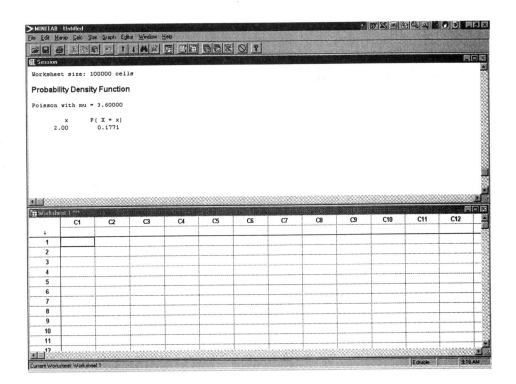

Technology Lab (pg. 185)

1. Create the Poisson distribution and save it in the Data Window. In C1, type in the values of X. Since n=20, the values of X are 0, 1, 2, 3, 4, 5, ...20. Next, use MINITAB to generate the Poisson probabilities for n=20 and μ=4. Click on **Calc → Probability Distributions → Poisson**. Select **Probability**. The **Mean** is 4. Now, tell MINITAB that the X values are in C1 and that you want the probabilities stored in C2. Enter C1 as the **Input Column** and enter C2 for **Optional Storage**. Click **OK**. The probabilities are in C2 of the Data Window. Notice, for example, that P(X=4) = .196367. This probability is the height on the histogram at X=4.

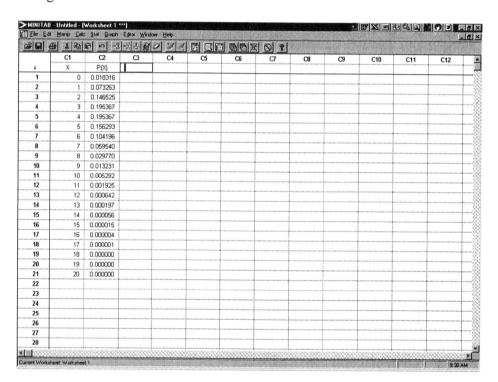

3. To generate 20 random numbers from a Poisson distribution with mean=4, click on **Calc → Random Data → Poisson**. Generate 20 **rows of data** and **Store in** C3. Enter a **Mean** of 4 and click on **OK**.

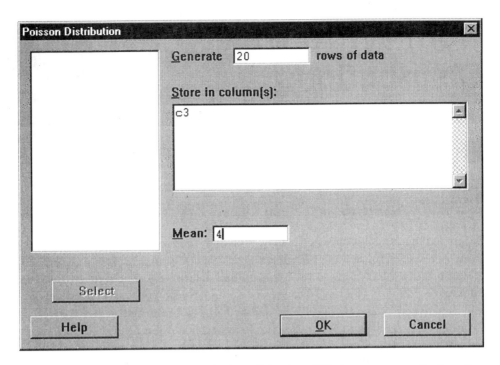

Use the random numbers that are in C3 of the Data Window to create (by hand) the table of waiting customers.

5. Repeat the steps in Exercise 3, but enter a **Mean** of 5 this time and **Store in** C4.
6. To calculate P(X=10) for a Poisson random variable with a mean of 5, click on **Calc → Probability Distributions → Poisson.** Since you want a simple probability, select **Probability** and enter 5 for the **Mean** and 10 for the **Input constant.**
7. To find the probabilities for parts a - c, use the Poisson probability distribution that you created in C1 and C2.

Normal Probability Distributions

CHAPTER 5

Section 5.1

▶ **Exercise 27 (pg. 201)** Graphing a normal curve with MINITAB

To graph the normal distribution with μ = 60 and σ = 12, you first have to store some X values in C1 and the corresponding probabilities in C2. Since most of the probability falls within 3σ of μ, this normal distribution should fall within the interval 60 ± 3(12), which is 24 to 96. To do this in MINITAB, you must put X values between 24 and 96 into C1. Click on **Calc → Make Patterned Data → Simple Set of Numbers.** You want to **Store patterned data in** C1. The X values should begin **From the first value** of 24 and go **To last value** of 96 **In steps of** 2. (Use steps of 2 because the interval from 24 to 96 is quite wide.)

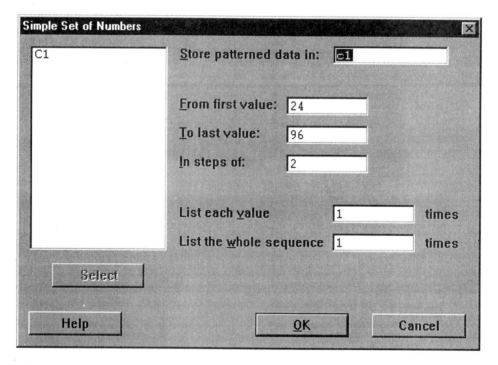

Click on **OK** and the numbers 24, 26, 28, ... 96 should be in C1. Next store the probabilities in C2. Click on **Calc → Probability Distribution → Normal.**

Section 5.1   75

Select **Probability density** and enter a **Mean** of 50 and a **Standard deviation** of 12. Enter C1 for the **Input column,** and C2 for the **Optional storage.**

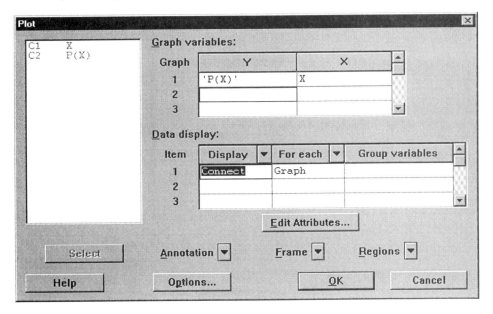

When you click on **OK**, the probabilities should be saved in C2. Label C1 as "X" and C2 as "P(X)". Now simply graph the distribution. Click on **Graph → Plot.** On the input screen, the **Y variables** are the probabilities contained in C2 and the **X variables** are the X-values contained in C1. Click on the down arrow to the right of **Display** and select **Connect.**

Click on **Annotation** → **Title** and enter an appropriate title. Click on **OK** to view the graph of the normal distribution with μ=60 and σ=12.

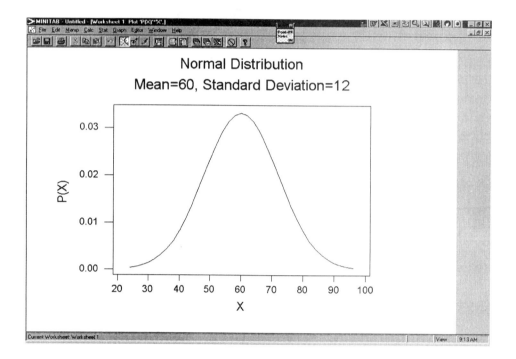

Section 5.3

> **Example 4 (pg. 215)** Using MINITAB to find Normal Probabilities.

Cholesterol levels of American men are normally distributed with μ=215 and σ=25. Find the probability that a randomly selected American man has a cholesterol level that is less than 175. To do this in MINITAB, click on **Calc → Probability Distributions → Normal.** On the input screen, select **Cumulative probability.** (Cumulative probability '*accumulates*' all probability to the left of

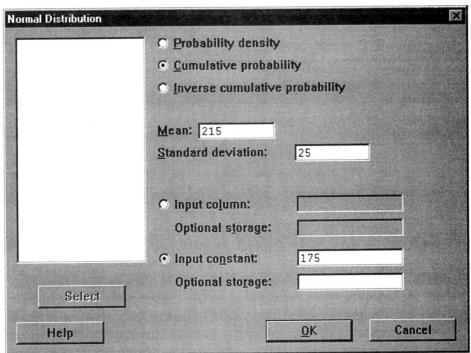

the input constant.) Enter 215 for the **Mean** and 25 for the **Standard deviation.** Next select **Input Constant** and enter the value 175.

Click on **OK** and the probability should be displayed in the Session Window. As you can see, the probability that a randomly selected American man has a cholesterol level that is less than 175 is equal to .0548.

Chapter 5 Normal Probability Distributions

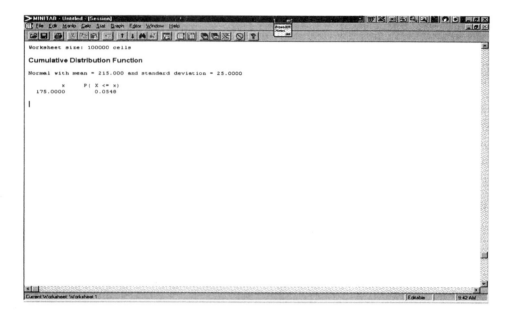

▶ **Example 5 (pg. 216)** Finding a specific data value

Scores for a civil service exam are normally distributed with μ=75 and σ=6.5. To be eligible for employment, you must score in the top 5%. Find the lowest score you can earn and still be eligible for employment. To do this in MINITAB, click on **Calc → Probability Distributions → Normal**. On the input screen, select **Inverse Cumulative probability.** Enter 75 for the **Mean** and 6.5 for the **Standard deviation.** For this type of problem, the **Input constant** will be the area to the left of the X-value we are looking for. This input constant will be a decimal number between 0 and 1. For this example, select **Input Constant** and enter the value .95 since 5% of the test scores are above this number and therefore, 95% are below this number. Click on **OK** and the X-value should be in the Session Window. Notice that the test score that qualifies you for employment is 85.6915.

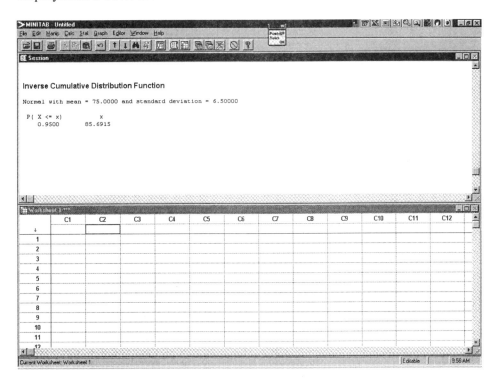

80 Chapter 5 Normal Probability Distributions

▶ Exercise 3 (pg. 217) Height of American Males

Heights are normally distributed with μ=69.2 inches and σ=2.9 inches. To complete parts (a) - (c), you will need MINITAB to give you two probabilities: one using X=66 and the other using X=72. Click on **Calc → Probability Distributions → Normal**. On the input screen, select **Cumulative probability**. Enter 69.2 for the **Mean** and 2.9 for the **Standard deviation**. Next select **Input Constant** and enter the value 66. Click on **OK**. Repeat the above steps using an **Input constant** of 72. Now the Session Window should have P(X ≤ 66) and P(X ≤ 72).

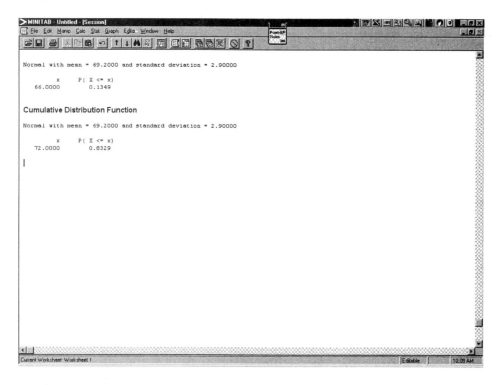

So, for Part (a), the P(X ≤ 66) = .1349. For Part (b), to find the P(66 ≤ X ≤ 72), you must subtract the two probabilities. Thus, the P(66 ≤ X ≤ 72) = .8329 - .1349 = .6980. For Part (c), to find P(X > 72), it is 1 - .8329 = .1671.

Section 5.4

▶ Example 6 (pg. 229) Finding Probabilities for X and $\overline{X}$

Credit card balances are normally distributed, with a mean of $2870 and a standard deviation of $900.

1. To find the probability that a randomly selected card holder has a balance less than $2500, click on **Calc → Probability Distributions → Normal**. On the input screen, select **Cumulative probability**. Enter 2870 for the **Mean** and 900 for the **Standard deviation**. Next select **Input Constant** and enter the value 2500. Click on **OK** and the probability should appear in the Session Window.

2. To find the probability that the *mean* balance of 25 card holders is less than $2500, you will need to calculate the standard deviation of $\overline{x}$ which is equal to $900 / \sqrt{25} = 180$. (Use a hand calculator for this calculation.) Now let MINITAB do the rest for you. Click on **Calc → Probability Distributions → Normal**. On the input screen, select **Cumulative probability**. Enter 2870 for the **Mean** and 180 for the **Standard deviation**. Next select **Input Constant** and enter the value 2500. Click on **OK** and the probability should appear in the Session Window.

◀

Technology Lab (pg. 234) Age Distribution in the United States

In this lab, you will compare the age distribution in the United States to the sampling distribution that is created by taking 36 random samples of size n=40 from the population and calculating the sample means.

Open worksheet **Tech5.** (Be sure that you have selected **Files of type** MINITAB Portable (MTP)). C1 should now contain the mean ages from the 36 random samples.

1. From the table on page 234 in your textbook, enter the Class Midpoints into C2. To enter the midpoints, click on **Calc → Make Patterned Data → Simple Set of Numbers.** On the input screen, you should **Store patterned data in** C2, **from the first value** of 2.5 **to the last value** of 97.5, **in steps of** 5. Click on **OK** and the midpoints should now be in C2. Next enter the relative frequencies, converted to proportions, into C3. So for a relative frequency of 7.3%, you will enter .073 into C3. The mean of this distribution is $\Sigma x\, p(x)$. To calculate the mean, you will have to multiply C2 and C3. To do this, click on **Calc → Calculator.** Type in the **Expression** C2 * C3 and **store result in variable** C4. Click on **OK** and in C4 you should now see the product of C2 and C3.

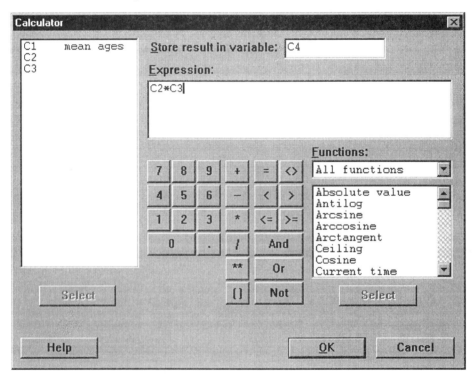

Now find the sum of C4. Click on **Calc → Column Statistics.** On the input screen, select **Sum** and use C4 for the **Input variable.** Click on **OK** and the column sum should be in the Session Window. As you can see, the mean age in the United States is 35.945.

2. The 36 sample means are in C1. To find the mean, click on **Stat → Basic Statistics → Display Descriptive Statistics.** Select C1 for the **Variable** and click on **OK**. The descriptive statistics will be displayed in the Session Window. The mean of the set of sample means is 36.209 and the standard deviation is 3.552. (You will need the standard deviation for question 6.)

4. To draw the histogram, click on **Graph → Histogram.** Select C1 for the **Graph variable.** In order to create a *relative frequency* histogram, click on **Options** and select **Percent.** Click on **OK** twice and you should be able to view the histogram.

5. To find the standard deviation of the ages of Americans, you must use the formula for the standard deviation of a Discrete Random variable, found on page 157 in the textbook. The shortcut formula will make this calculation easier. Use the formula $\Sigma x^2 p(x) - \mu^2$ and take the square root of this value. In MINITAB, first square all the midpoints. Click on **Calc → Calculator.** Type in the **Expression** C2 * C2 and **store result in variable** C5. Click on **OK** and in C5 you should now see the midpoints squared. To calculate $x^2 p(x)$, you must multiply C5 by C3. Click on **Calc → Calculator.** Type in the **Expression** C5 * C3 and **store result in variable** C6.

84 Chapter 5 Normal Probability Distributions

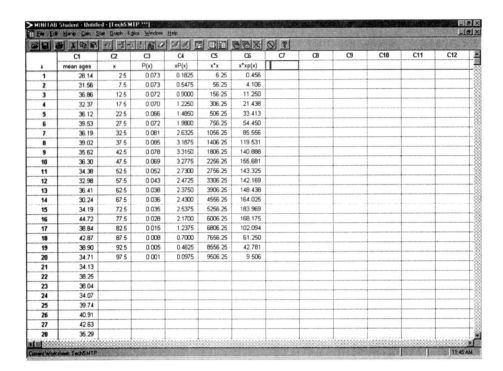

Now find the sum of C6. Click on **Calc → Column Statistics**. On the input screen, select **Sum** and use C6 for the **Input variable**. Click on **OK** and the column sum should be in the Session Window. As you can see, $\Sigma x^2 p(x)$ is 1792.5. Next, subtract μ^2 from 1792.5 and take the square root. (Recall that $\mu = 35.945$) Click on **Calc → Calculator**. Type in the **Expression** SQRT(1792.5 -35.945*35.945) and **store result in variable** C7. The standard deviation is the number now in C7.

6. The standard deviation of the 36 sample means can be found in the descriptive statistics that you produced for question 2.

Confidence Intervals

CHAPTER 6

Section 6.1

▶ **Example 4(pg. 256)** Construct a 99% Confidence Interval for the mean number of sentences in an ad

Open the worksheet **Sentence** which is found in the **Chapter 6** Folder. The data will be in C1. First find the standard deviation of the data. Click on **Calc → Column Statistics**. Select **Standard deviation** and enter C1 for the **Input variable**. Click on **OK** and the standard deviation will be displayed in the Session Window. To construct the confidence interval, click on **Stat → Basic Statistics → 1-Sample Z**. Select C1 for the **Variable**. Next, select **Confidence Interval** and enter 99.0 for the **Level**. For **Sigma,** enter the standard deviation that is displayed in the Session Window.

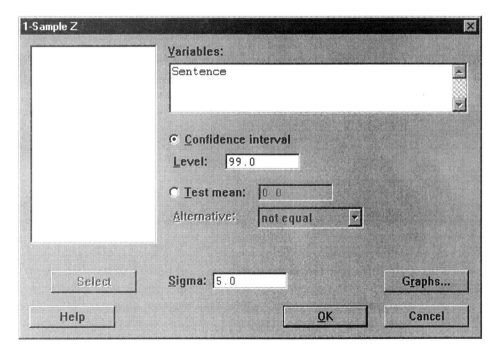

Click on **OK** and the results will be displayed in the Session Window.

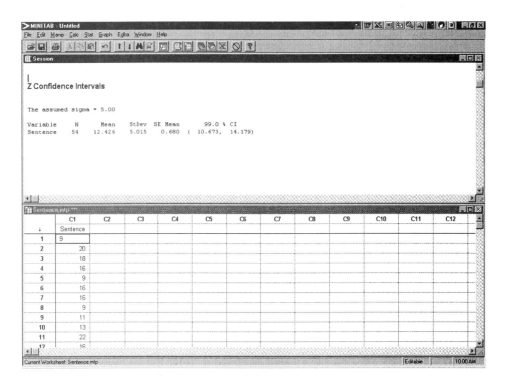

As you can see, the 99% confidence interval is (10.673, 14.179).

Section 6.1   87

▶ **Exercise 31 (pg. 261)** Construct 90% and 99% confidence intervals for mean reading time

Enter the data into C1. To construct the confidence interval, click on **Stat → Basic Statistics → 1-Sample Z.** Select C1 for the **Variable**. Next, select **Confidence Interval** and enter 90.0 for the **Level.** For **Sigma,** enter the assumed value of 1.5. Click on **OK** and the interval will be displayed in the Session Window. Repeat the above steps, but use 99.0 for the confidence **Level.**

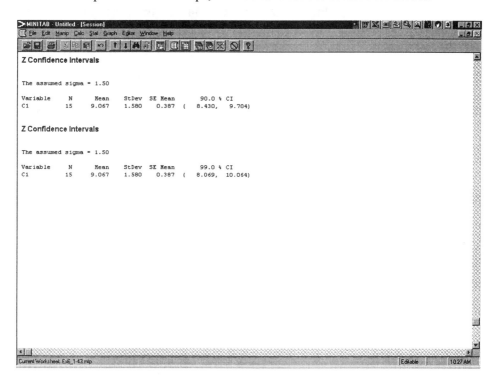

Exercise 43 (pg. 263) Construct a 95% confidence interval for airfare prices

Open worksheet **Ex6_1-43.mtp** in the **Chapter 6** Folder. The airfares are in C1. Find the standard deviation of the sample. Click on **Calc → Column Statistics.** Select **Standard deviation** and enter C1 for the **Input variable.** Click on **OK** and the standard deviation will be displayed in the Session Window. To construct the confidence interval, click on **Stat → Basic Statistics → 1-Sample Z.** Select C1 for the **Variable.** Next, select **Confidence Interval** and enter 95.0 for the **Level.** For **Sigma,** enter the standard deviation that is displayed in the Session Window.

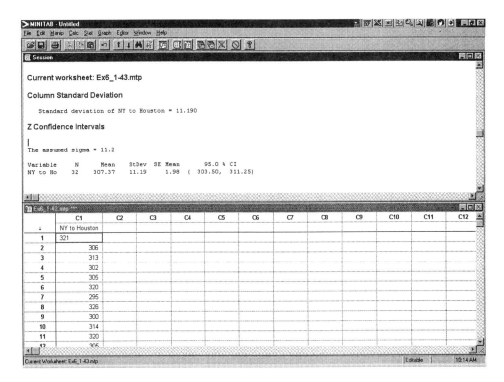

Section 6.2

▶ **Example 2 (pg. 268)** Construct a 95% Confidence Interval for the mean temperature of coffee sold

Enter the temperatures of the coffee sold at 16 randomly selected restaurants. The data is found in the middle of page 290 of the textbook. Since n=16 and the population standard deviation is unknown, you should construct a t-interval for this problem. Click on **Stat → Basic Statistics → 1-Sample t.** Select C1 for the **Variable**. Next, select **Confidence Interval** and enter 95.0 for the **Level.** Click on **OK** and the output will be displayed in the Session Window.

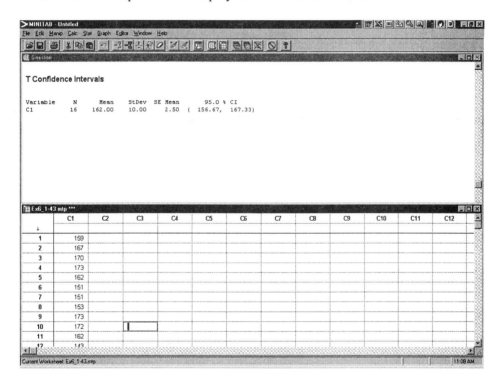

Exercise 19 (pg. 272) Construct a 99% confidence interval for the mean SAT scores

Enter the SAT scores into C1. Since n=12 and the population standard deviation is unknown, you should construct a t-interval for this problem. Click on **Stat → Basic Statistics → 1-Sample t.** Select C1 for the **Variable**. Next, select **Confidence Interval** and enter 99.0 for the **Level**. Click on **OK** and the output will be displayed in the Session Window.

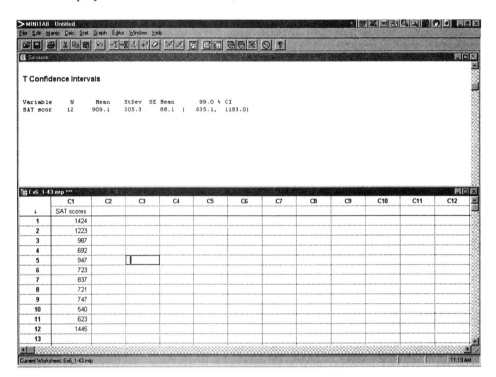

Section 6.3

▶ **Example 2 (pg. 277)** Construct a 95% confidence interval for p

From Example 1, on page 275 of the textbook, you know that 380 of 883 American adults said that their favorite sport was football. To construct a 95% confidence interval, click on on **Stat → Basic Statistics → 1 Proportion.** Select **Summarized Data.** The **Number of trials** is 883 and the **Number of successes** is 380.

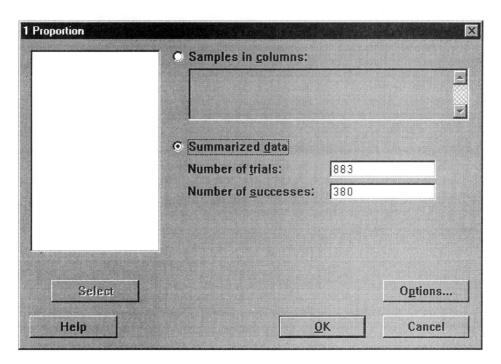

Next, to select the confidence level, click on **Options.** Enter 95.0 for the **Confidence Interval.** Also, select **Use test and interval based on normal distribution.**

92 Chapter 6 Confidence Intervals

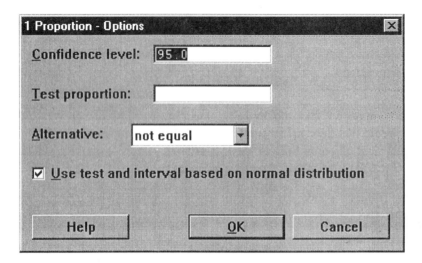

Click on **OK** twice and the output will be displayed in the Session Window.

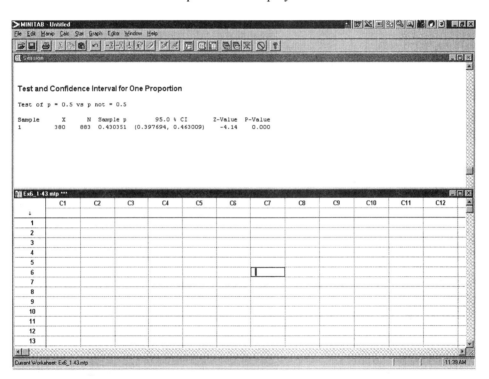

Notice the interval is (.397694, .463009).

▶ **Exercise 11 (pg. 280)** Construct intervals for the proportion of adults with low levels of iodine

A study of 34000 American adults found that 4080 had low levels of iodine. Construct both 95% and 99% confidence intervals for the true proportion of adults with low levels of iodine. Click on **Stat → Basic Statistics → 1 Proportion**. Select **Summarized Data**. The **Number of trials** is 34000 and the **Number of successes** is 4080. Next, to select the confidence level, click on **Options**. Enter 95.0 for the **Confidence Interval**. Also, select **Use test and interval based on normal distribution**. Click on **OK** twice and the results will be in the Session Window. Repeat using 99.0 for the **Confidence Interval**.

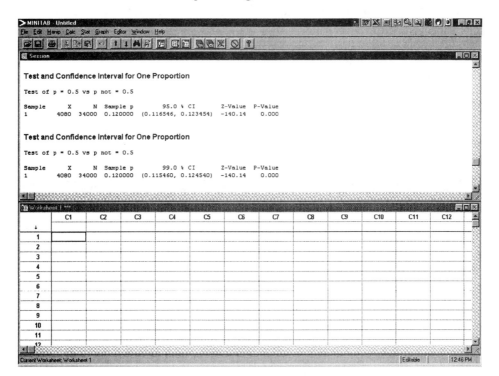

Technology Lab (pg. 283)

1. Click on **Stat → Basic Statistics → 1 Proportion**. Select **Summarized Data**. The **Number of trials** is 1005 and the **Number of successes** is 141. Next, to select the confidence level, click on **Options**. Enter 95.0 for the **Confidence Interval**. Also, select **Use test and interval based on normal distribution**. Click on **OK** twice and the results will be in the Session Window.

3. Oprah Winfrey was named by 6% of the people in the sample. That means that 60 of the 1005 named Oprah Winfrey. Use the steps in question 1 to construct the 95% confidence interval. This time the **Number of successes** is 60.

4. To do this simulation, you will generate random binomial data with n=1005 and p=.07. The result displayed in each cell of C1 will represent the number of people who named Oprah. Click on **Calc → Random Data → Binomial**. **Generate** 200 **rows of data** and **Store in column** C1. The **Number of trials** is 1005 and the **Probability of success** is .07. When you click on **OK**, C1 will contain a simulation of 200 samples.

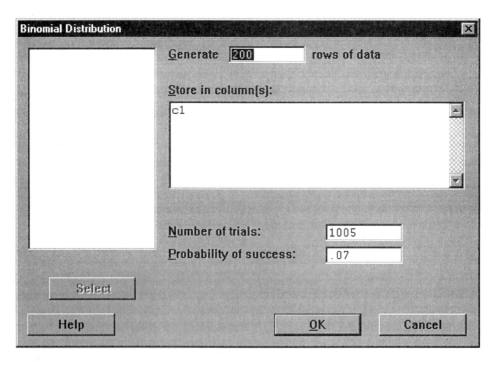

To calculate the sample proportions, click on **Calc → Calculator**. Enter C1/1005 for the **Expression** and **Store the result in** C2. Click on **OK.** Now C2 contains 200 values of the sample proportion. Sort C2 to find the smallest and largest value. Click on **Manip** → **Sort**. **Sort column** C2 and **Store sorted column in** C3. You want to **Sort by column** C2.

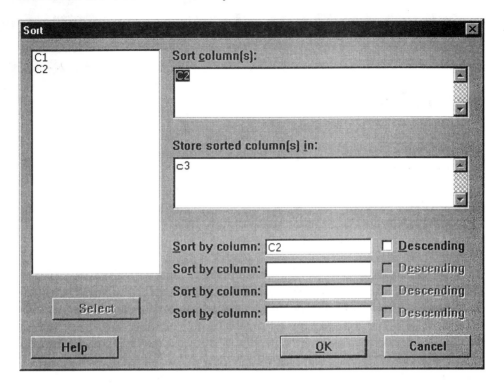

Click on **OK** and C3 should contain the sample proportions sorted from smallest to largest. Thus, the smallest value should be in Row 1 and the largest value should be in Row 200.

Hypothesis Testing with One Sample

Section 7.2

▶ **Example 9 (pg. 323)** Hypothesis Testing Using P-values

You think that the average franchise investment information given in the graph on page 323 of the textbook is incorrect, so you randomly select 30 franchises and determine the necessary investment for each. Is there enough evidence to support your claim at $\alpha = .05$? Use the P-value to interpret.

Open worksheet **Franchise,** which is found in the **Chapter 7** MINITAB folder. The data should be in C1. To do a 1-Sample Z-test in MINITAB, you must know σ. Click on **Calc → Column Statistics.** Select **Standard deviation** for the **Statistic** to be calculated and enter C1 for the **Input Variable.** Click on **OK** and the standard deviation will be in the Session Window. You will enter this value in the input screen for the 1-Sample Z test. Click on **Stat → Basic Statistics → 1-Sample Z.** Enter C1 for the **Variable,** click on **Test mean** and enter 143260. Since the claim is "the mean is different from $143,260", you will perform a two-tailed test. **Alternative** should be "not equal". Beside **Sigma**, enter the value that you found for σ, 30000.

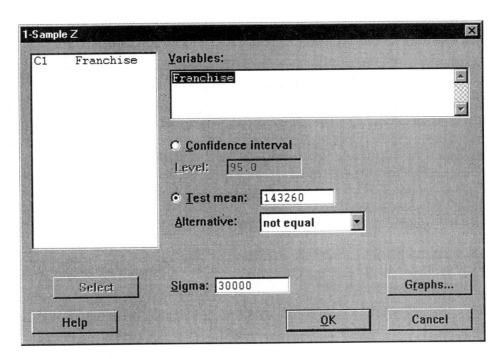

Click on **OK** and the results should be displayed in the Session Window.

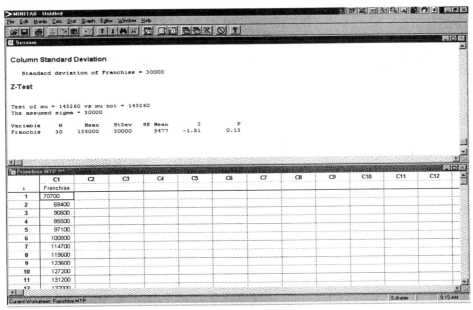

Notice that both the test statistic and the P-value are given. From the output, note that z = -1.51 and P = .13. Since the P-value is larger than α, you should Fail to Reject the null hypothesis.

Exercise 25 (pg. 326) Nitrogen Dioxide Level in West London

Open worksheet **Ex7_2-25**, which is found in the **Chapter 7** MINITAB folder. Click on **Calc → Column Statistics**. Select **Standard deviation** for the **Statistic** to be calculated and enter C1 for the **Input Variable**. Click on **OK** and the standard deviation will be in the Session Window. You will enter this value in the input screen for the 1-Sample Z test. Click on **Stat → Basic Statistics → 1-Sample Z.** Enter C1 for the **Variable,** click on **Test mean** and enter 28. Since the claim is "the mean is greater than 28 parts per billion", you will perform a right-tailed test. Click on the down arrow beside **Alternative** and select "greater than". Beside **Sigma**, enter the value that you found for σ, 22.128. Click on **OK** and the results of the test should be displayed in the Session Window.

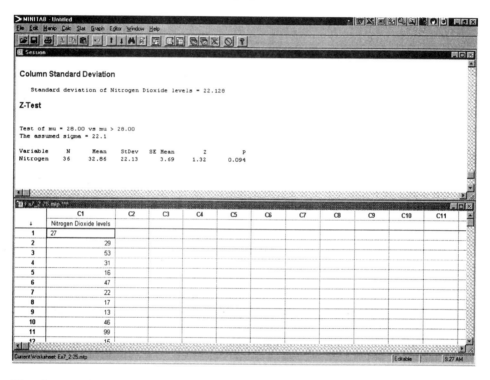

Use the P-value (.094) to draw a conclusion about the scientist's estimate at the level of significance α = .05.

▶ **Exercise 33 (pg. 328)** Years taken to quit smoking permanently

Open worksheet **Ex7_2-33,** which is found in the **Chapter 7** MINITAB folder. Click on **Calc → Column Statistics.** Select **Standard deviation** for the **Statistic** to be calculated and enter C1 for the **Input Variable.** Click on **OK** and the standard deviation will be in the Session Window. You will enter this value in the input screen for the 1-Sample Z test. Click on **Stat → Basic Statistics → 1-Sample Z.** Enter C1 for the **Variable,** click on **Test mean** and enter 15. Since the claim is "the mean time is 15 years", you will perform a two-tailed test. Click on the down arrow beside **Alternative** and select "not equal". Beside **Sigma,** enter the value that you found for σ, 4.2876. Click on **OK** and the results of the test should be displayed in the Session Window.

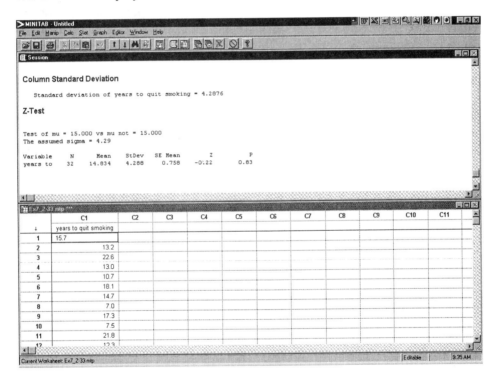

Section 7.3

Example 4 (pg. 333) Testing μ with a Small Sample

A used car dealer says that the mean price of a 1995 Ford F-150 Super Cab is at least $16,500. You suspect this claim is incorrect. At α = .05, is there enough evidence to reject the dealer's claim?

Enter the data into C1. (The data can be found on page 358 of the textbook.) Be sure to enter the data in numeric form. So enter the amount $14,500 as 14500. Since this is a small sample problem, you will be performing a 1-Sample t-test. For this type of problem, MINITAB calculates the sample standard deviation automatically. Click on **Stat → Basic Statistics → 1-Sample t.** Enter C1 for the **Variable,** click on **Test mean** and enter 16500. Since you suspect that the used car dealer's claim is incorrect, you will perform a left-tailed test. Click on the down arrow beside **Alternative** and select "less than". Click on **OK** and the results of the test should be displayed in the Session Window.

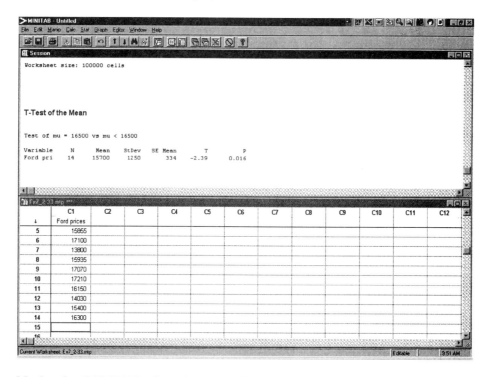

Notice that MINITAB gives the test statistic and the P-value, so that you can make your conclusion using either value. Since the P-value is smaller than α, you should Reject the null hypothesis.

Section 7.3   101

▶ Exercise 23 (pg. 338) Soda consumed by teen-age males

Open worksheet **Ex7_3-23,** which is found in the **Chapter 7** MINITAB folder. Click on **Stat → Basic Statistics →1-Sample t.** Enter C1 for the **Variable,** click on **Test mean** and enter 3. Since the claim is "teenage males drink less than 3 servings", you will perform a left-tailed test. Click on the down arrow beside **Alternative** and select "less than". Click on **OK** and the results of the test should be displayed in the Session Window.

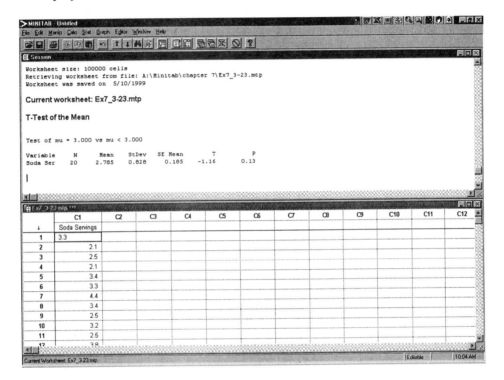

102 Chapter 7 Hypothesis Testing with One Sample

▶ Exercise 24 (pg. 338) Water consumed by American adults

Open worksheet **Ex7_3-24**, which is found in the **Chapter 7** MINITAB folder. Click on **Stat → Basic Statistics → 1-Sample t**. Enter C1 for the **Variable**, click on **Test mean** and enter 5. Since the claim is "American adults drink less than 5 8-oz. glasses of water per day", you will perform a left-tailed test. Click on the down arrow beside **Alternative** and select "less than". Click on **OK** and the results of the test should be displayed in the Session Window.

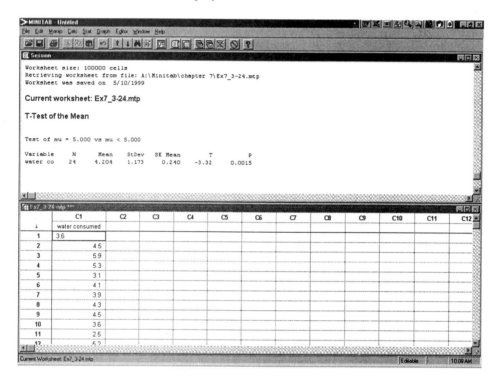

Section 7.4

▶ Example 2 (pg. 342) Hypothesis Test for a Proportion

Of 200 Americans, 27% are in favor of outlawing cigarettes. At $\alpha = .05$, is there enough evidence to reject the claim that 23% of Americans favor outlawing cigarettes?

Click on **Stat → Basic Statistics → 1-Proportion**. The data is given in a summarized form, so select **Summarized data**. Enter 200 for the **Number of trials**. Since 27% of the sample were in favor, the **Number of Successes** is 54 (.27 * 200).

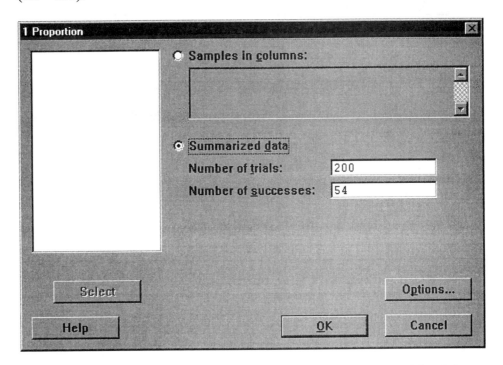

Click on **Options**. Enter .23 for the **Test Proportion** because it is claimed that 23% are in favor, and select "not equal" for the **Alternative**. Since np and nq are both larger than 5, click on **Use test and interval based on normal distribution**, and then click on **OK** twice.

104 Chapter 7 Hypothesis Testing with One Sample

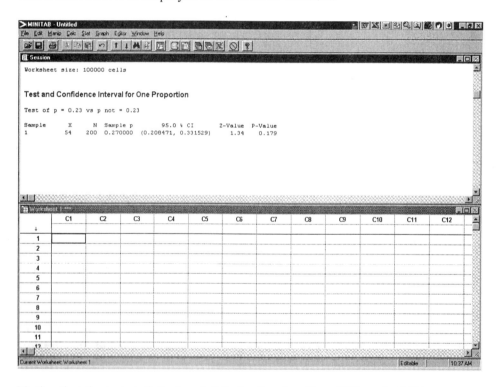

The results should be displayed in the Session Window.

Notice that the test statistic (z = 1.34), the P-value (P = .179) and a 95% confidence interval for the true proportion of Americans in favor of outlawing cigarettes are all displayed in the output. With such a large P-value, you should Fail to Reject the null hypothesis.

Section 7.4   105

▸ Exercise 9 (pg. 344) Consumer Pollution Concerns

Click on **Stat → Basic Statistics → 1-Proportion.** The data is given in a summarized form, so select **Summarized data**. Enter 1050 for the **Number of trials.** Since 32% of the sample have stopped buying this product because of pollution concerns, the **Number of Successes** is 336 (.32 * 1050 = 336). Click on **Options**. Enter .30 for the **Test Proportion** because it is claimed that more than 30% have stopped buying the product, and select "greater than" for the **Alternative.** Click on **Use test and interval based on normal distribution,** and then click on **OK** twice.

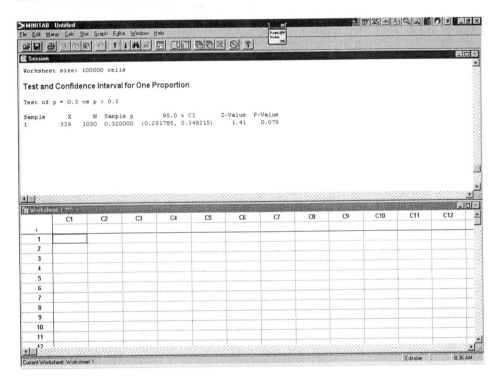

Exercise 11 (pg. 345) Government Regulation of Business

Click on **Stat → Basic Statistics → 1-Proportion**. The data is given in a summarized form, so select **Summarized data**. Enter 1762 for the **Number of trials**. Since 1004 Americans in the sample believe that regulation does more harm than good, the **Number of Successes** is 1004. Click on **Options**. Enter .60 for the **Test Proportion** because it is claimed that 60% have this view, and select "not equal" for the **Alternative**. Click on **Use test and interval based on normal distribution**, and then click on **OK** twice.

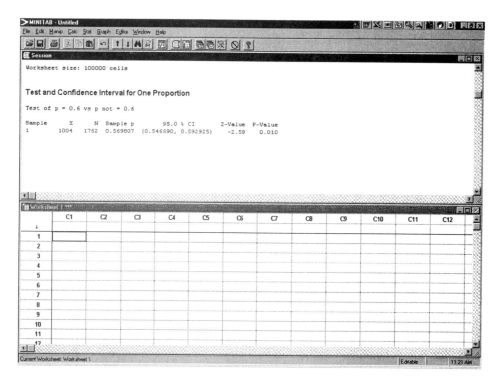

Technology Lab (pg. 346) The Case of the Vanishing Women

4. Click on **Stat → Basic Statistics → 1-Proportion.** The data is given in a summarized form, so select **Summarized data**. Enter 100 for the **Number of trials.** Since 9 women were selected, the **Number of Successes** is 9. Click on **Options**. Enter .2914 for the **Test Proportion** because 29.14% of the original sample were women, and select "not equal" for the **Alternative**. Click on **Use test and interval based on normal distribution,** and then click on **OK** twice.

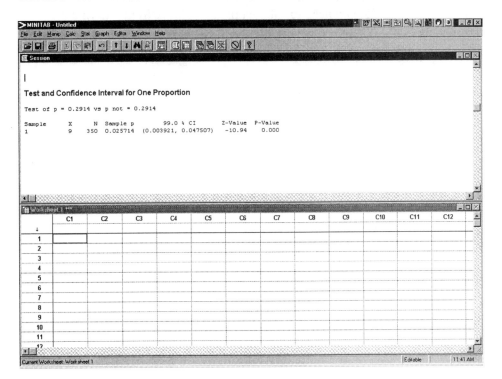

Extending the Basics (pg. 355) — Finding P-values for a Chi-Square Test

Once you have calculated the test statistic, χ_0^2, MINITAB can calculate the exact P-value of the test for you. For example, suppose $\chi_0^2 = 55.758$ and there are 40 degrees of freedom. Click on **Calc → Probability distributions → Chi-square.** Select **Cumulative Probability** and enter 40 **Degrees of Freedom.** Next, click on **Input Constant** and enter the test statistic, 55.758. Click on **OK** and the output will be displayed in the Session Window. Since you have found $P(\chi_0^2 < 55.758) = .95$, you need to subtract the probability from 1 to find the P-value. So, the P-value is $1 - .95 = .05$.

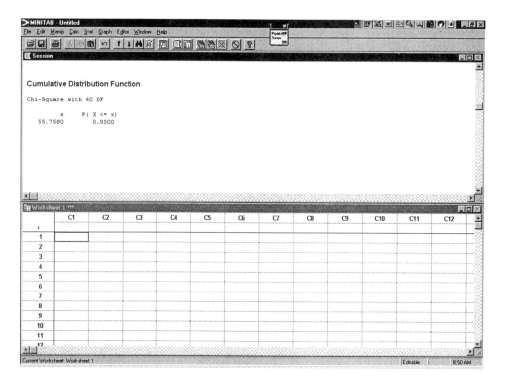

Hypothesis Testing with Two Samples

CHAPTER 8

Section 8.1

▶ **Exercise 13 (pg. 375)** Time spent watching TV

Open worksheet **Ex8_1-13.mtp**, which is found in the **Chapter 8** MINITAB folder. The 1981 data (Time A) is in C1 and the new data (Time B) is in C2. Notice that for both samples, n = 30. MINITAB does not have a 2-sample Z-test, but you can use a 2-sample t-test instead since the t distribution becomes very similar to the normal distribution as the sample size approaches 30. Click on **Stat → Basic Statistics → 2-Sample t**. Select **Samples in different columns** and enter C1 for the **First** and C2 for the **Second** column. Click on the down arrow beside **Alternative** and select **greater than** since the sociologist claims that children spent more time watching TV in 1981 than they do today.

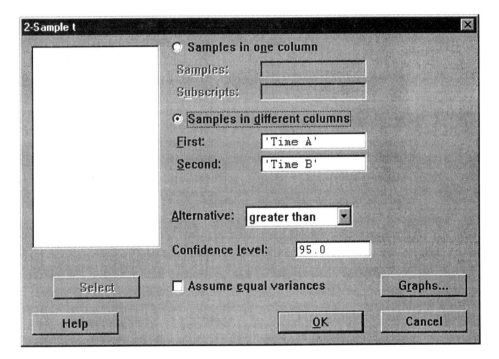

Click on **OK** and the results of the test should be displayed in the Session Window.

Chapter 8 Hypothesis Testing with Two Samples

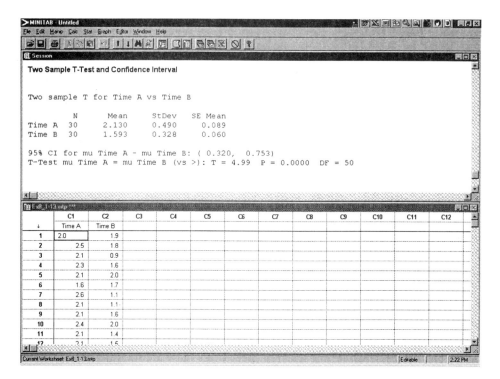

Notice that the test statistic is T = 4.99 with a P-value = .0000. Since this P-value is so small, you would Reject H_0 at any α level.

Section 8.1 111

> **Exercise 15 (pg. 375)** Difference between Washer Diameters

Open worksheet **Ex8_1-15.mtp**, which is found in the **Chapter 8** MINITAB folder. The diameters from the first method are in C1 and the diameters from the second method are in C2. Notice that for both samples, n = 35. MINITAB does not have a 2-sample Z-test, but you can use a 2-sample t-test instead since the t distribution becomes very similar to the normal distribution as the sample size gets larger than 30. Click on **Stat → Basic Statistics → 2-Sample t.** Select **Samples in different columns** and enter C1 for the **First** and C2 for the **Second** column. Click on the down arrow beside **Alternative** and select **not equal** since the production engineer claims there is no difference between the two methods. Click on **OK** and the results of the test should be displayed in the Session Window.

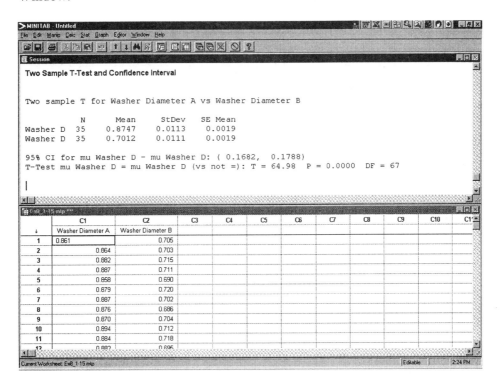

Since the P-value is so small, you would Reject H_0 at any α level.

112 Chapter 8 Hypothesis Testing with Two Samples

Section 8.2

> **Example 1 (pg. 382)** Snow Tire Performance

Enter the data, found on page 408 of the textbook, into the MINITAB Data Window. Put the Firestone data in C1 and the Michelin data in C2. Click on **Stat → Basic Statistics → 2-Sample t**. Select **Samples in different columns** and enter C1 for the **First** and C2 for the **Second** column. Click on the down arrow beside **Alternative** and select **not equal** since you want to test whether the mean stopping distances are different. Click on **OK** and the results of the test should be displayed in the Session Window.

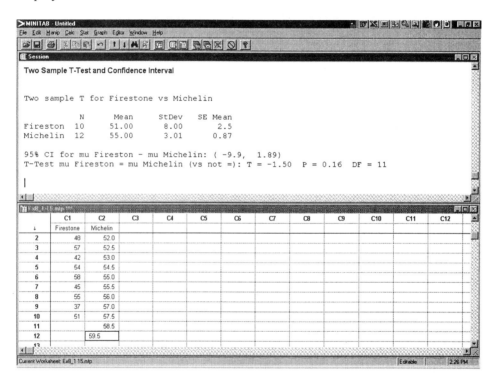

Notice that the test statistic is T= -1.5. Since the P-value = .16 and is greater than the α-level of .01, there is not enough evidence to conclude that the mean stopping distances of the tires are different.

▸ Exercise 17 (pg. 386) Tensile Strength of Steel Bars

Enter the data, found on page 386 of the textbook, into the MINITAB Data Window. Put the New method data in C1 and the Old method data in C2. Click on **Stat → Basic Statistics → 2-Sample t.** Select **Samples in different columns** and enter C1 for the **First** and C2 for the **Second** column. Click on the down arrow beside **Alternative** and select **not equal** since you want to test if the new treatment makes a difference in the tensile strength of steel bars. Select **Assume Equal Variances,** since the problem tells you to assume the population variances are equal. Click on **OK** and the results of the test should be displayed in the Session Window.

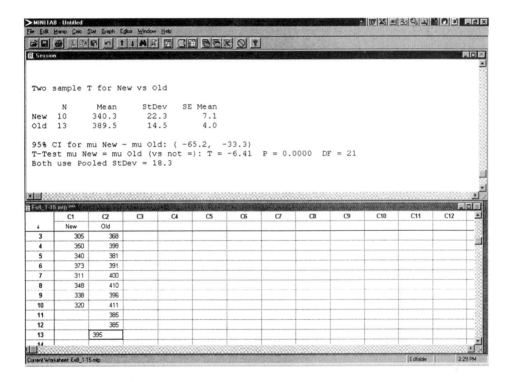

114 Chapter 8 Hypothesis Testing with Two Samples

Exercise 20 (pg. 387) Comparing Teaching Methods

Enter the data, found on page 387 of the textbook, into the MINITAB Data Window. Put the Traditional Lab data in C1 and the Interactive data in C2. Click on **Stat → Basic Statistics → 2-Sample t.** Select **Samples in different columns** and enter C1 for the **First** and C2 for the **Second** column. Click on the down arrow beside **Alternative** and select **less than** since you want to test if the students taught in a traditional lab had lower science test scores than students taught with the interactive simulation software. Select **Assume Equal Variances,** since the problem tells you to assume the population variances are equal. Click on **OK** and the results of the test should be displayed in the Session Window.

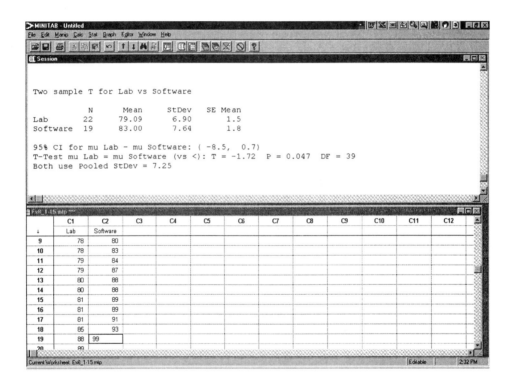

Section 8.3

▶ Example 2 (pg. 392) Golf Scores

Enter the data, found on page 392 of the textbook, into the MINITAB Data Window. Put the Old Design data in C1 and the New Design data in C2. Click on **Stat → Basic Statistics → Paired t.** Enter C1 for the **First Sample** and C2 for the **Second Sample.**

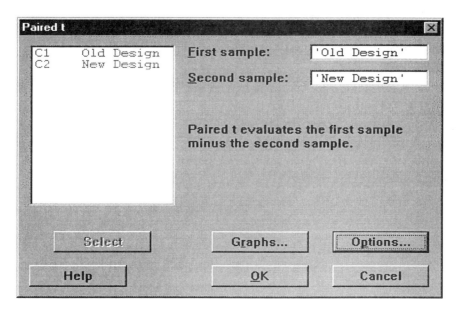

Click on **Options.** Enter 0 for **Test Mean** and select **greater than** as the **Alternative.** Click on **OK** twice to display the results.

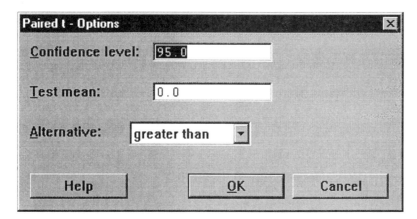

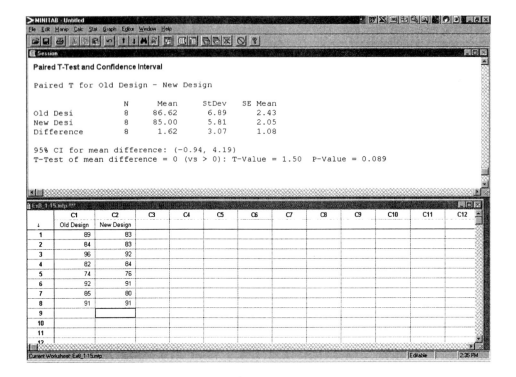

Section 8.3   117

▸ Exercise 16 (pg. 396) Verbal SAT Scores

Enter the data, found on page 396 of the textbook, into the MINITAB Data Window. Put the scores on the first SAT in C1 and the scores on the second SAT in C2. Click on **Stat → Basic Statistics → Paired t.** Enter C1 for the **First Sample** and C2 for the **Second Sample.** Click on **Options.** Enter 0 for **Test Mean** and select **less than** as the **Alternative** because, if the scores have improved, then the difference (first SAT - second SAT) will be less than 0. Click on **OK** twice to display the results.

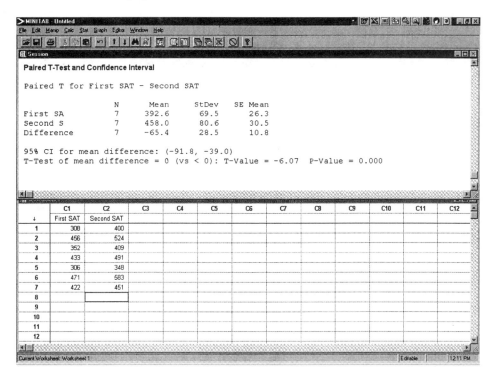

118   Chapter 8   Hypothesis Testing with Two Samples

Exercise 17 (pg. 397) Does a Fuel Additive Improve Mileage?

Enter the data, found on page 397 of the textbook, into the MINITAB Data Window. Put the scores on the mileage without additive in C1 and the mileage with additive in C2. Click on **Stat → Basic Statistics → Paired t**. Enter C1 for the **First Sample** and C2 for the **Second Sample**. Click on **Options**. Enter 0 for **Test Mean** and select **less than** as the **Alternative** because if the mileage are improved with the additive, then the difference (without additive - with additive) will be less than 0. Click on **OK** twice to display the results.

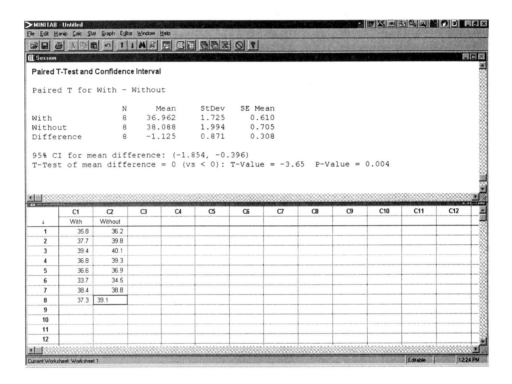

Since the P-value is so small, you should Reject the null hypothesis. Thus, the additive appears to improve mileage.

Section 8.4

▶ **Example 1 (pg. 402)** Difference between male and female Internet Users

In a study of 200 female and 250 male Internet users, 30% of the females and 38% of the males plan to shop on-line. This is a summary of the results of the study. To test if there is a difference in the proportion of male and female users who plan to shop on-line, click on **Stat → Basic Statistics → 2 Proportions**. Select **Summarized Data** and use the data for Females as the **First sample**. Enter 200 **Trials** and 60 **Successes** (200 x .30). Use the data for Males as the **Second sample**. Enter 250 **Trials** and 95 **Successes** (250 x .38).

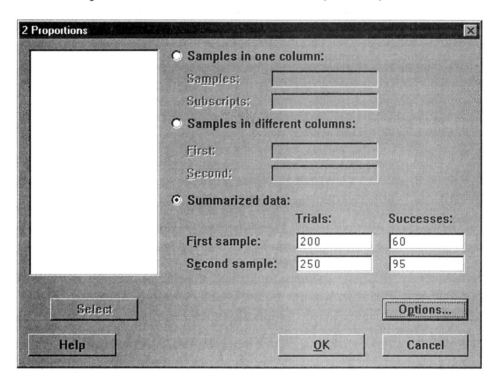

Click on **Options**. Enter 0 for **Test mean**, and select **not equal** as the **Alternative** since you want to test if there is a difference between the proportion of male and female shoppers. Next click on **Use pooled estimate of p for test.**

120 Chapter 8 Hypothesis Testing with Two Samples

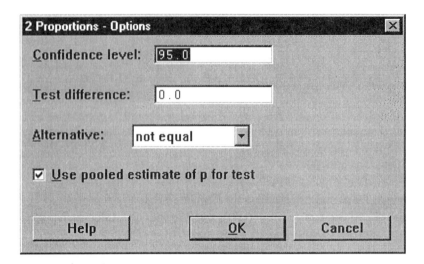

Click on **OK** twice to display the results in the Session Window.

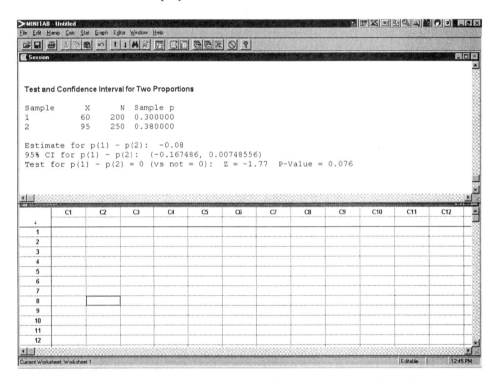

Since the P-value is smaller than α, you should Reject the null hypothesis.

Exercise 7 (pg. 404) Alternative Medicine Usage

To test if there is a difference in the proportion of adults who used alternative medicine in 1990 and the proportion of adults who use it now, click on **Stat → Basic Statistics → 2 Proportions**. Select **Summarized Data** and use the data for 1990 as the **First sample**. Enter 1539 **Trials** and 520 **Successes**. Use the recent study data as the **Second sample**. Enter 2055 **Trials** and 865 **Successes**. Click on **Options**. Enter 0 for **Test mean**, and select **not equal** as the **Alternative** since you want to test if there is a difference between the proportion of users in 1990 and the present. Next click on **Use pooled estimate of p for test**. Click on **OK** twice to display the results in the Session Window.

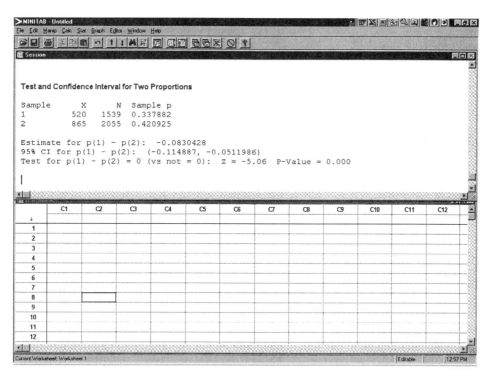

With such a small P-value, you should Reject the null hypothesis. Thus, the proportion of adults using alternative medicines has changed since 1990.

Exercise 10 (pg. 405) Fewer Smokers in California?

To test if the proportion of adult smokers in California is lower than the proportion of adult smokers in Oregon, click on **Stat → Basic Statistics → 2 Proportions**. Select **Summarized Data** and use the data for California as the **First sample**. Enter 1500 **Trials** and 276 **Successes** (1500 x .184). Use the Oregon data as the **Second sample**. Enter 100 **Trials** and 207 **Successes**. Click on **Options**. Enter 0 for **Test mean**, and select **less than** as the **Alternative** since you want to test if the proportion of smokers in California is lower than the proportion in Oregon. Next click on **Use pooled estimate of p for test**. Click on **OK** twice to display the results in the Session Window.

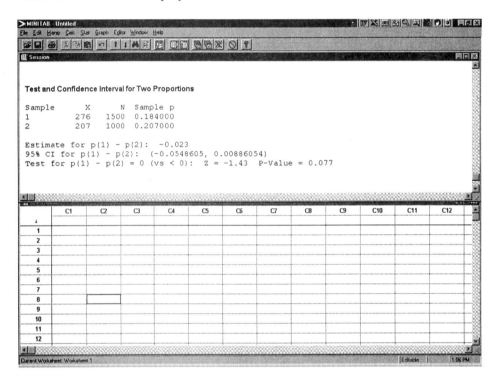

Technology Lab (pg. 407) Tails Over Heads

1. Click on **Stat → Basic Statistics → 1 Proportion**. Select **Summarized Data** and enter 11902 **Trials** and 5772 **Successes**. Click on **Options**. Enter .5 for **Test proportion**, and select **not equal** as the **Alternative** since you want to test if the probability is .5 or not. Next click on **Use test and interval based on normal distribution**. Click on **OK** twice to display the results in the Session Window.

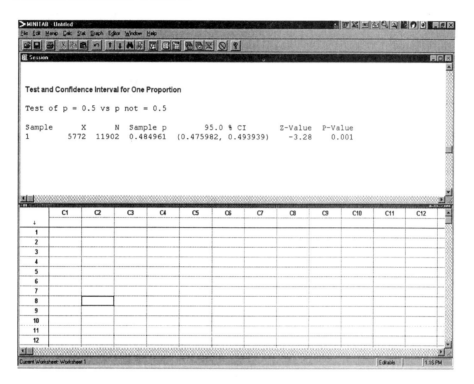

3. To repeat the simulation, click on **Calc → Random data → Binomial**. You want to **Generate** 500 **rows of data** and Store In Column C1. The **Number of trials** is 11902 and the **Probability of success** is .5. Click on **OK** and C1 should have 500 rows of data in it. To draw the histogram, click on **Graph → Histogram**. Enter C1 below **X** and beside **Graph 1**. Click on **OK** to view the histogram.

Correlation and Regression

CHAPTER 9

Section 9.1

Example 3 (pg. 420) Constructing a Scatter Plot

Open worksheet **OldFaithful,** which is found in the **Chapter 9** MINITAB folder. The duration (in minutes) of several of Old Faithful's eruptions should be in C1, and the time (in minutes) until the next eruption should be in C2. Notice that Duration is the x-variable and Time is the y-variable. To plot the data, click on **Graph → Plot.** On the input screen, enter C2 for the **Y variable** and C1 for the **X variable.**

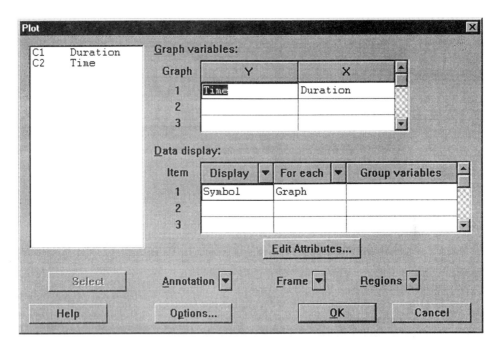

Next, click on **Frame → Tick.** Set the tick mark positions so that both axes begin at zero. On the top half of the input screen, below **Positions,** enter the tick positions for the **X variable** as 0:5/.5 (0 to 5 in steps of .5) and for the **Y variable** as 0:100/10 (0 to 100 in steps of 10).

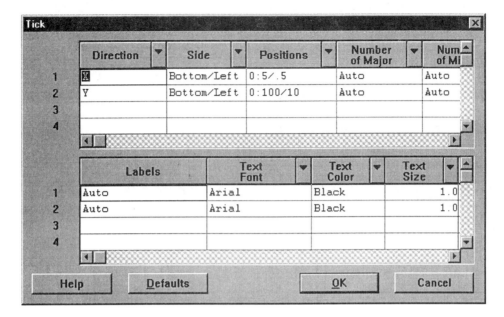

Click on **OK** twice to view the scatter plot.

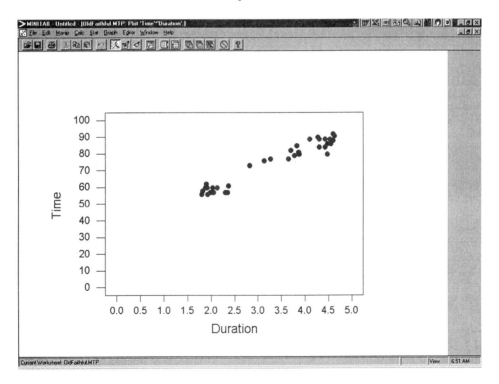

126 Chapter 9 Correlation and Regression

▶ Example 5 (pg. 423) Finding the Correlation Coefficient

Open worksheet **OldFaithful,** which is found in the **Chapter 9** MINITAB folder. The duration (in minutes) of several of Old Faithful's eruptions should be in C1, and the time (in minutes) until the next eruption should be in C2. Notice that Duration is the x-variable and Time is the y-variable. To find the correlation coefficient, click on **Stat → Basics Statistics → Correlation.** On the input screen, select both C1 and C2 for **Variables,** by double-clicking on each one.

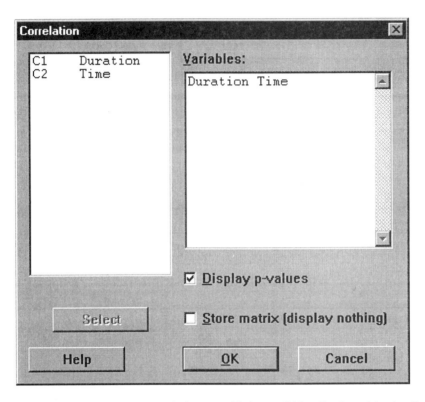

Click on **OK** and the Correlation Coefficient will be displayed in the Session Window.

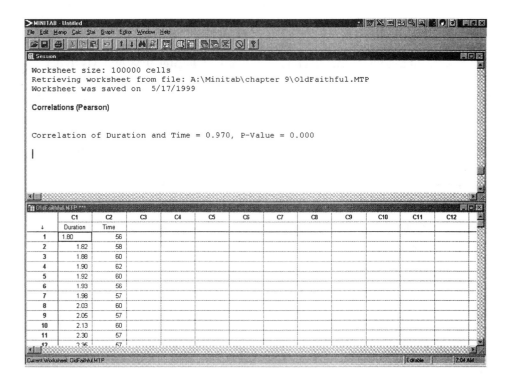

128 Chapter 9 Correlation and Regression

▶ Exercise 13 (pg. 428) Plot of Study Hours vs. Test Scores

Open worksheet **Ex9_1-13**, which is found in the **Chapter 9** MINITAB folder. The hours spent studying should be in C1, and the test scores should be in C2. Notice that "Hours" is the x-variable and "Test Scores" is the y-variable. To plot the data, click on **Graph → Plot**. On the input screen, enter C2 for the **Y variable** and C1 for the **X variable**. Next, click on **Frame → Tick**. Set the tick mark positions so that both axes begin at zero. On the top half of the input screen, below **Positions**, enter the tick positions for the **X variable** as 0:8/1 (0 to 8 in steps of 1) and for the **Y variable** as 0:100/10 (0 to 100 in steps of 10). Click on **OK** twice to view the scatter plot.

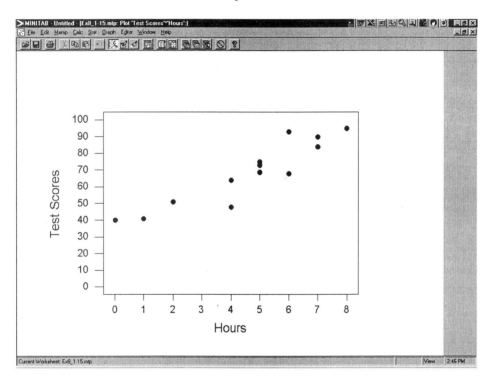

Now, to find the correlation coefficient, click on **Stat → Basics Statistics → Correlation**. On the input screen, select both C1 and C2 for **Variables**, by double-clicking on each one. Click on **OK** and the output should be in the Session Window.

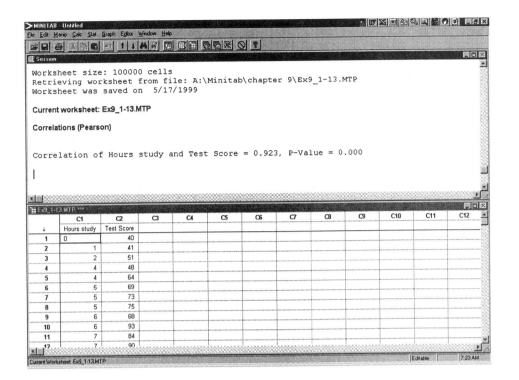

Section 9.2

▶ Example 2 (pg. 434) Finding a Regression Equation

Open worksheet **OldFaithful,** which is found in the **Chapter 9** MINITAB folder. The duration (in minutes) of several of Old Faithful's eruptions should be in C1, and the time (in minutes) until the next eruption should be in C2. Notice that Duration is the x-variable and Time is the y-variable. To find the regression equation, click on **Stat → Regression → Regression.** Enter C2 for the **Response** variable, and C1 as the **Predictor.**

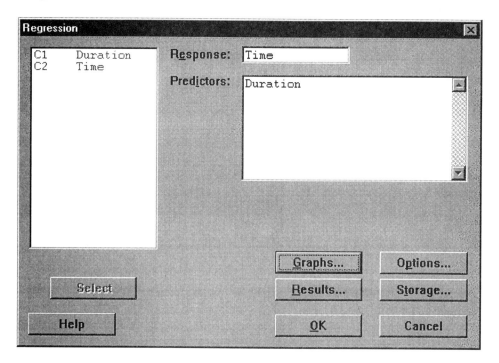

Click on **Results.** Select **Regression equation, table of coefficients, s, R-squared, and basic analysis of variance.**

Section 9.2   131

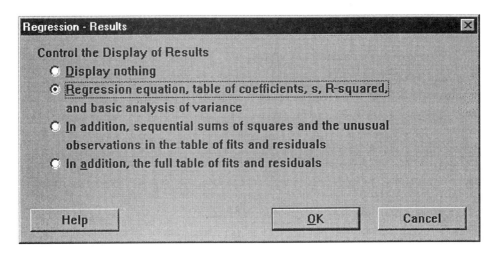

Click on **OK** twice to view the output in the Session Window.

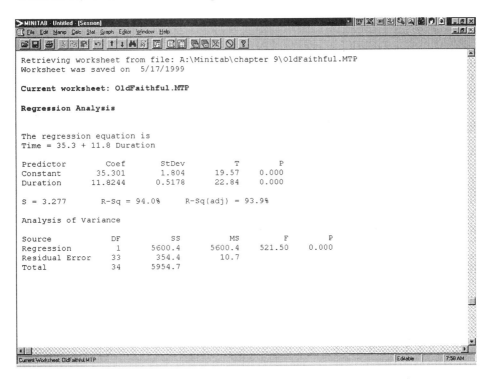

Notice that the regression equation is Time = 35.3 + 11.8 * Duration.

132 Chapter 9 Correlation and Regression

▶ Exercise 11 (pg. 437) Age vs. Systolic blood pressure

Enter the data into the MINITAB Data Window. Put Age into C1 and Blood Pressure into C2. First plot the data. Click on **Graph → Plot**. On the input screen, enter C2 for the **Y variable** and C1 for the **X variable**. Click on **OK**.

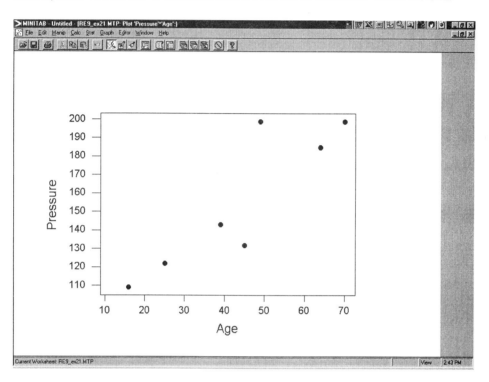

To find the regression equation, click on **Stat → Regression → Regression**. Enter C2 for the **Response** variable, and C1 as the **Predictor**. Click on **Results**. Select **Regression equation, table of coefficients, s, R-squared, and basic analysis of variance**. Click on **OK** twice.

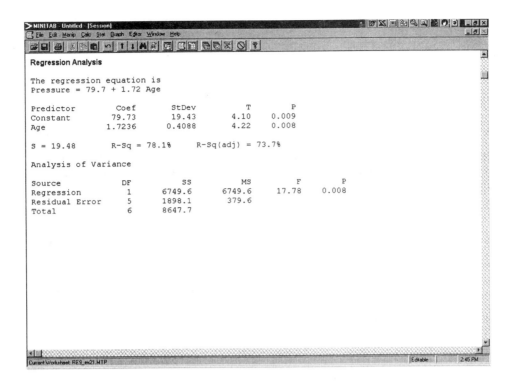

Section 9.3

> **Example 2 (pg. 445)** Finding the Standard Error and the Coefficient of Determination

Enter the first two columns of data (found on page 445 of the textbook) into the MINITAB Data Window. Enter the x's into C1 and name it Expenses. Enter the y's into C2 and name it Sales. Both the coefficient of determination and the standard error of the estimate are part of the regression output. To find the regression equation, click on **Stat → Regression → Regression**. Enter C2 for the **Response** variable, and C1 as the **Predictor**. Click on **Results**. Select **Regression equation, table of coefficients, s, R-squared, and basic analysis of variance.** Click on **OK** twice.

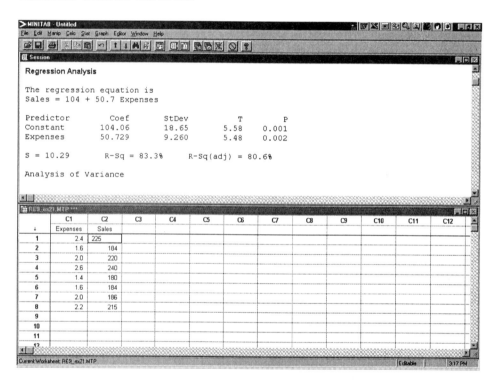

Notice that the standard error of the estimate is S = 10.29 and the coefficient of determination is R-Sq = 83.3%.

> **Example 3 (pg. 447)** Constructing a Prediction Interval

Enter the first two columns of data (found on page 445 of the textbook) into the MINITAB Data Window. Enter the x's into C1 and name it Expenses. Enter the y's into C2 and name it Sales. To find the regression equation, click on **Stat → Regression → Regression.** Enter C2 for the **Response** variable, and C1 as the **Predictor.** Now, to find both the point estimate and the prediction interval, click on **Options. (Fit Intercept** is selected by default.) Next enter the advertising expenditure. Although the amount given in this problem is $2100, the data is stored in thousands of dollars. So enter 2.1 for **Prediction interval for new observations.** Enter 95 for the **Confidence level** and select **Prediction limits.**

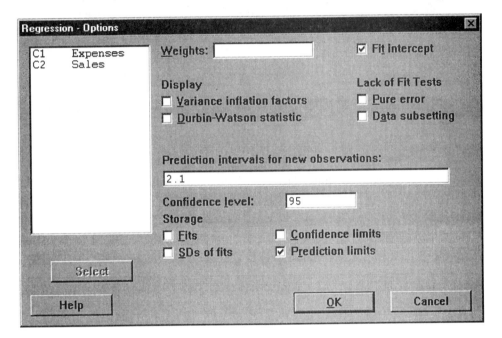

Click on **OK**. Click on **Results.** If you would like to see the other regression output, then select **Regression equation, table of coefficients, s, R-squared, and basic analysis of variance.** If you only want to see the prediction interval, then select **Display nothing.** Click on **OK** twice and view the output in the Session Window.

136 Chapter 9 Correlation and Regression

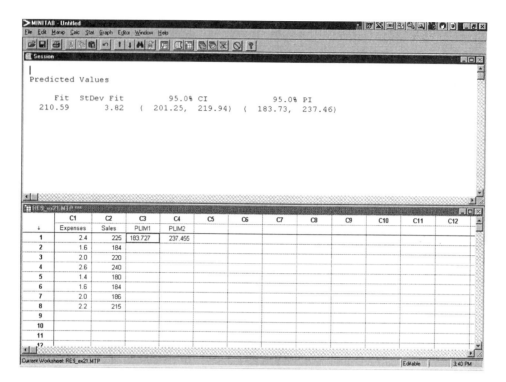

Notice that the predicted value, 210.59, is listed below **Fit** and the prediction interval, (183.73, 237.46), is listed below **95% PI**.

Exercise 11 (pg. 449) and Exercise 19 (pg. 450) Retail space vs. Sales

Open worksheet **Ex9_3-11,** which is found in the **Chapter 9** MINITAB folder. Square footage is in C1 and Sales is in C2. For these two problems, you need to find the coefficient of determination, the standard error of the estimate, and a 90% prediction interval when square footage is 4.5 billion. This can be accomplished all at once in MINITAB. Click on **Stat → Regression → Regression.** Enter C2 for the **Response** variable, and C1 as the **Predictor.** Now, to find both the point estimate and the prediction interval, click on **Options.** Select **Fit Intercept** by clicking on it. Next enter the square footage. So enter 4.5 for **Prediction interval for new observations.** Enter 90 for the **Confidence level** and select **Prediction limits.** Click on **OK**. Click on **Results.** Since you would like to see the other regression output, then select **Regression equation, table of coefficients, s, R-squared, and basic analysis of variance.** Click on **OK** twice and view the output in the Session Window.

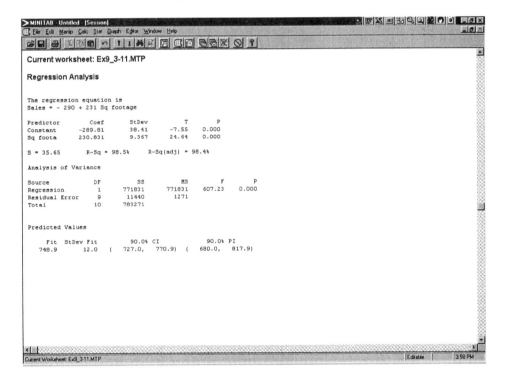

Section 9.4

> **Example 1 (pg. 452)** Finding a Multiple Regression Equation

Open worksheet **Salary,** which is found in the **Chapter 9** MINITAB folder. Salary should be in C1, Employment in C2, Experience in C3, and Education in C4. Click on **Stat → Regression → Regression.** Enter C1 for the **Response** variable, and enter C2, C3, and C4 as the **Predictors.** Click on **Results.** Since you would like to see the other regression output, then select **Regression equation, table of coefficients, s, R-squared, and basic analysis of variance.** Click on **OK** twice and view the output in the Session Window.

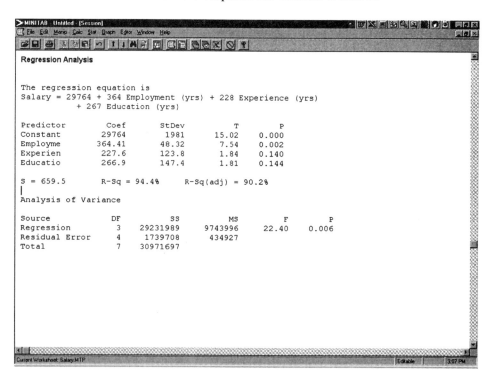

The regression equation is listed at the beginning of the output. Notice that the regression equation uses the values listed below **Coef.** These values are the coefficients of the multiple regression equation.

Exercise 5 (pg. 456) Finding a Multiple Regression Equation

Open worksheet **Ex9_4-5,** which is found in the **Chapter 9** MINITAB folder. Sales should be in C1, Square footage in C2, and Number of shopping centers in C3. Click on **Stat → Regression → Regression**. Enter C1 for the **Response** variable, and enter C2 and C3 as the **Predictors.** Click on **Results.** Since you would like to see the other regression output, then select **Regression equation, table of coefficients, s, R-squared, and basic analysis of variance.** Click on **OK** twice and view the output in the Session Window.

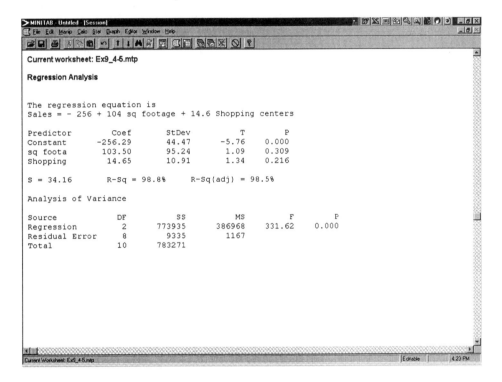

140 Chapter 9 Correlation and Regression

▶ Technology Lab (pg. 457) Tar, Nicotine, and Carbon Monoxide

Open worksheet **Tech9,** which is found in the **Chapter 9** MINITAB folder. Brand is in C1, Tar is in C2, Nicotine is in C3, Weight is in C4, and Carbon Monoxide is in C5.

1. Click on **Graph → Plot.** On the input screen, enter C2 for the **X variable** and C3 for the **Y variable.** Click on **OK.** Repeat these steps for parts b - f, changing the variables as directed in each part.

3. Click on **Stat → Basics Statistics → Correlation.** On the input screen, select both C2 and C3 for **Variables,** by double-clicking on each one. Click on **OK** and the output should be in the Session Window. Repeat these steps for each pair of variables listed in question 1, parts b - f.

4. (Do both 4 & 5 at one time here). Click on **Stat → Regression → Regression.** Enter C3 for the **Response** variable, and enter C2 as the **Predictor.** Next, to predict the nicotine content of a cigarette with 13 mg. of tar, click on **Options.** Select **Fit Intercept** by clicking on it. Next enter the tar content. So enter 13 for **Prediction interval for new observations.** Enter 95 for the **Confidence level** and select **Prediction limits.** Click on **OK.** Next, click on **Results.** Since you would like to see the other regression output, select **Regression equation, table of coefficients, s, R-squared, and basic analysis of variance.** Click on **OK** twice and view the output in the Session Window. Repeat these steps using C5 for the **Response** variable.

6. (Do both 6 & 7 at one time here). Click on **Stat → Regression → Regression.** Enter C2 for the **Response** variable, and enter C3, C4, and C5 as the **Predictors.** Next, click on **Results.** Since you would like to see the other regression output, then select **Regression equation, table of coefficients, s, R-squared, and basic analysis of variance.** Click on **OK** twice and view the output in the Session Window. For part b, repeat these steps using C3 and C5 for the **Predictors**. Next, to predict the tar content of a cigarette with 1 mg. of nicotine and 10 mg. of carbon monoxide, click on **Options.** Select **Fit Intercept** by clicking on it. Next enter the nicotine and carbon monoxide contents. To enter both of these, type in a 1 (for the nicotine), leave a space, and then type in a 10 (for the carbon monoxide) for **Prediction interval for new observations.** Click on **OK.** Next, click on **Results.** Since you would like to see the other regression output, then select **Regression equation, table of coefficients, s, R-squared, and basic analysis of variance.** Click on **OK** twice and view the output in the Session Window.

Chi-Square Tests and the F-Distribution

CHAPTER 10

Section 10.1

> Example 3 (pg. 470)   The Chi-Square Goodness-of-Fit Test

Enter the data into the MINITAB Data Window. Enter the Responses into C1 and name it Response. Enter the frequencies into C2 and name it Observed. Enter the distribution (from the null hypothesis) into C3 and name it Distribution. These values should be proportions, not percentages. Thus, 61% should be entered as .61.

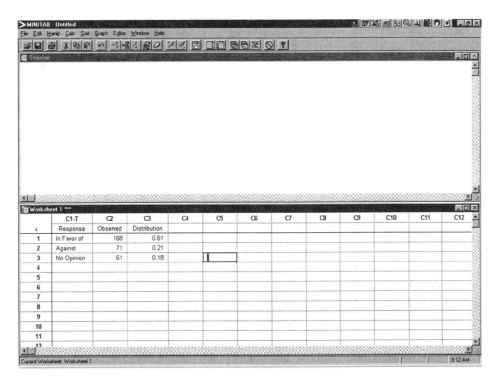

To calculate the expected frequencies, you will multiply the distribution times the sample size, which is 300 in this example. Click on **Calc → Calculator.** You will **Store the result in** C4, and calculate the **Expression** C3*300. Click on **OK.** Name C4 Expected since it now contains the expected frequencies.

142 Chapter 10 Chi-Square Tests and the F-Distribution

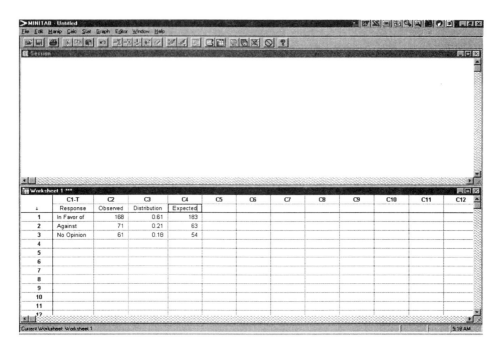

Next, calculate the chi-square test statistic, $(O - E)^2 / E$. Click on **Calc → Calculator.** You will **Store the result in** C5, and calculate the **Expression** (C2 - C4)**2 / C4. Click on **OK** and C5 should contain the calculated values.

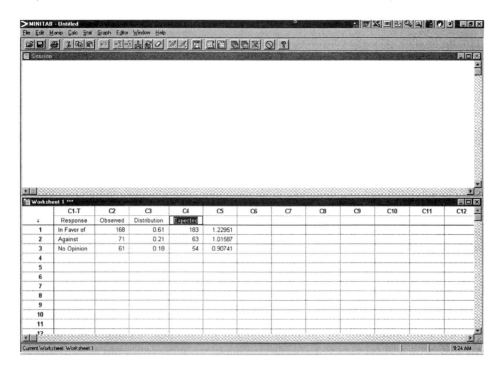

Now, just add up the values in C5 and the sum is the test statistic. Click on **Calc → Column Statistics.** Select **Sum** and enter C5 for the **Input Variable.** Click on **OK** and the Chi-Square test statistic will be displayed in the Session Window. In this example, the test statistic is 3.1528. Next, calculate the P-value to help you decide if you should reject the null hypothesis. Click on **Calc → Probability Distributions → Chi-square.** Select **Cumulative Probability** and enter 2 **Degrees of Freedom.** Enter the value of the test statistic, 3.1528, for the **Input Constant.** Click on **OK** and $P(X \leq 3.1528)$ will be in the Session Window.

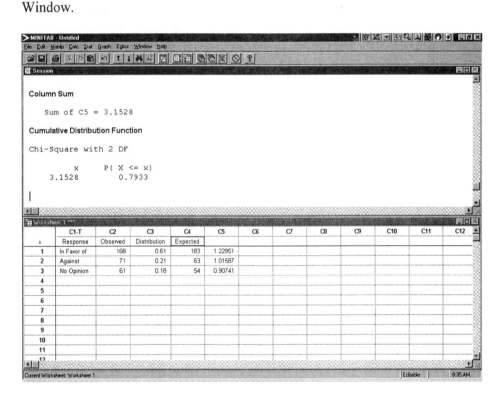

The $P(X \leq 3.1528)$ is .7933. The P-value is $P(X \geq 3.1528)$, which is $1 - P(X \leq 3.1528)$. This value is $1 - .7933 = .2067$. Since this P-value is larger than $\alpha = .05$, you should not reject the null hypothesis.

Exercise 5 (pg. 473) Roadside Hazard Crash Deaths

Enter the data into the MINITAB Data Window. Enter the Objects into C1 and name it Object. Enter the frequencies into C2 and name it Observed. Enter the distribution into C3 and name it Distribution. (These values should be entered as proportions, and not percentages.) To calculate the expected frequencies, you will multiply the distribution times the sample size, which is 691 in this example. The sample size can be found by adding up the frequencies. (To do this in MINITAB, click on **Calc → Column Statistics.** Select **Sum** and enter C2 for the **Input Variable**.) Click on **Calc → Calculator.** You will **Store the result in** C4, and calculate the **Expression** C3*691. Click on **OK**. Name C4 Expected since it now contains the expected frequencies. Next, calculate the chi-square test statistic, $(O - E)^2 / E$. Click on **Calc → Calculator.** You will **Store the result in** C5, and calculate the **Expression** (C2 - C4)**2 / C4. Click on **OK** and C5 should contain the calculated values.

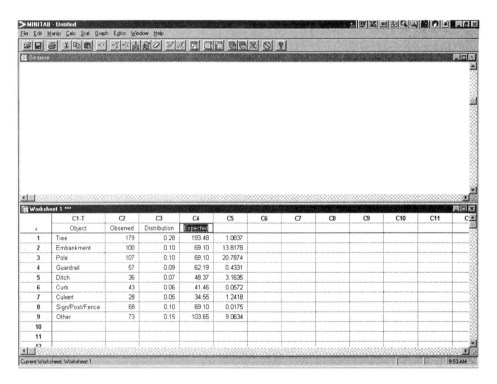

Now, just add up the values in C5 and the sum is the test statistic. Click on **Calc → Column Statistics.** Select **Sum** and enter C5 for the **Input Variable**. Click on **OK** and the Chi-Square test statistic will be displayed in the Session Window. In this example, the test statistic is 49.665.

Next, calculate the P-value to help you decide if you should reject the null hypothesis. Click on **Calc → Probability Distributions → Chi-square.** Select **Cumulative Probability** and enter 8 **Degrees of Freedom.** Enter the value of the test statistic, 49.665, for the **Input Constant.** When you click on **OK**, the $P(X \leq 49.665)$ will be in the Session Window. The P-value is $P(X \geq 49.665) = 1 - P(X \leq 49.665) = 1 - 1 = 0$. Since this P-value is so small, you would reject the null hypothesis at any α-level.

146 Chapter 10 Chi-Square Tests and the F-Distribution

Section 10.2

▶ Example 3 (pg. 481) Chi-Square Independence Test

Enter the data into the MINITAB Data Window. First label the columns: use Gender for C1, "0-1" for C2, ... "6-7" for C5. Now enter the data into the appropriate columns. Do not enter any totals.

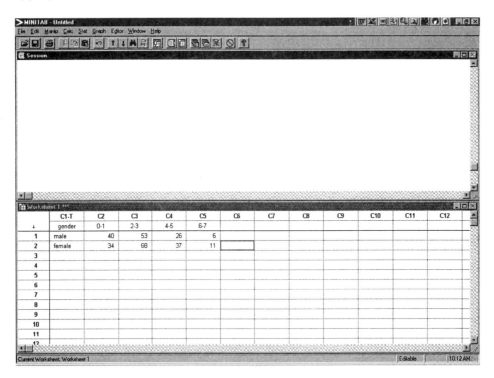

To perform the chi-square independence test, click on **Stat → Tables → Chi-square Test.** On the input screen, select C2 - C5 for the **Columns containing the table.** Click on **OK** and the test results will be displayed in the Session Window.

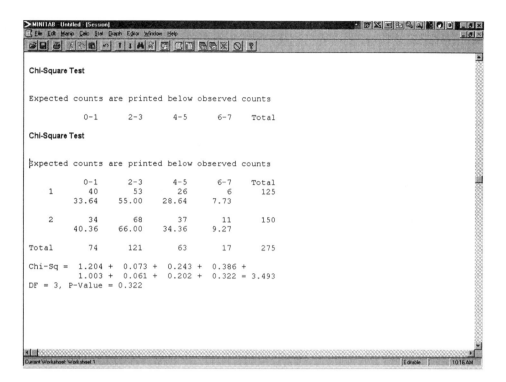

Notice that the test statistic is Chi-Sq = 3.493 and the P-value = .322. Since this P-value is larger than $\alpha = .05$, you should not reject the null hypothesis. Thus, there is not enough evidence to conclude that the number of days per week spent exercising is related to gender.

Exercise 5 (pg. 483) Should the drug be used as treatment?

Enter the data into the MINITAB Data Window. Enter Result data into C1, Drug data into C2 and Placebo data into C3. To perform the chi-square independence test, click on **Stat → Tables → Chi-square Test**. On the input screen, select C2 - C3 for the **Columns containing the table**. Click on **OK** and the test results will be displayed in the Session Window.

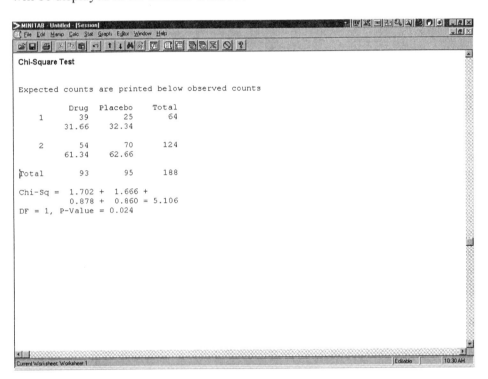

Notice that the test statistic is Chi-Sq = 5.106 and the P-value = .024. Since this P-value is smaller than α = .10, you should reject the null hypothesis. Thus, there is evidence to conclude that the drug should be used as part of the treatment.

Section 10.3

> **Picturing the World (pg. 491)** Performing a Two-Sample F-Test

MINITAB requires the raw data values in order to perform a two-sample F-test, so you will use the data "Picturing the World" found in the margin of page 491. Enter the data for the Northeast section into C1 and for the Other sections into C2. To perform a two-sample F-test, MINITAB requires that the data be stacked into one column with a second column identifying which sample each data value came from. To do this, click on **Manip** → **Stack/Unstack** → **Stack Columns**. Select both columns to be stacked on top of each other. **Store stacked data in** C3 and **Store subscripts in** C4. The subscripts will be numbers 1 or 2 to indicate which column the data value came from.

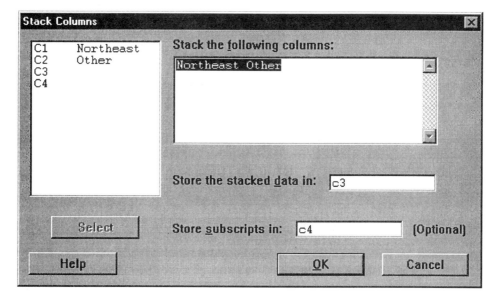

Click on **OK** and look in the Data Window. C3 should have all of the data stored in it and C4 should have the numbers 1 (this tells you that the data value came from C1) or 2 (this tells you that the data value came from C2) in it.

150 Chapter 10 Chi-Square Tests and the F-Distribution

Now, to perform the two-sample F-test, click on **Stat → ANOVA → Homogeneity of Variance.** On the input screen, C3 is the **Response** variable and C4 is the **Factor.** Enter an appropriate **Title** and click on **OK.** This test produces quite a lot of output. However, you are only interested in the results of the F-test.

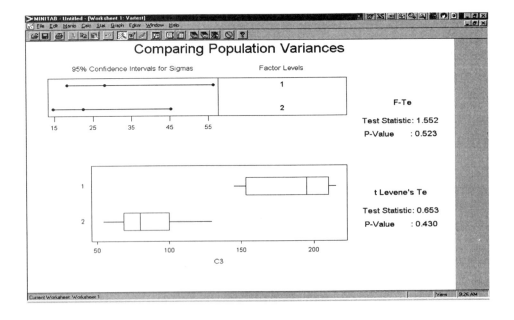

You can simply close the Graph Window, and look at the results in the Session Window.

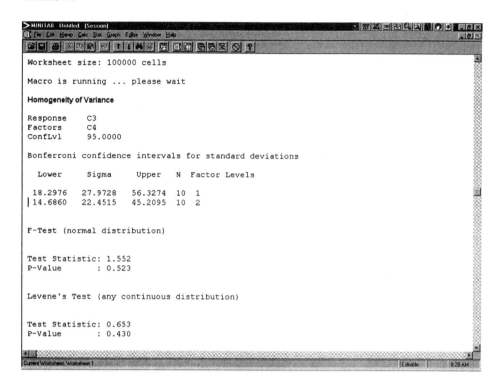

The F-test results show that the test statistic = 1.552 and the P-value = .523. Since this is a large P-value, you would fail to reject the null hypothesis. Thus, the two variances are approximately equal.

Section 10.4

> **Example 2 (pg. 501)** ANOVA Tests

Open worksheet **Airline** found in the MINITAB folder **Chapter 10**. The data for the three airlines should be in C1, C2, and C3. To perform a one-way analysis of variance, click on **Stat → ANOVA → One Way (Unstacked)**. Select all three columns and click on **OK**. The results of the test will be in the Session Window.

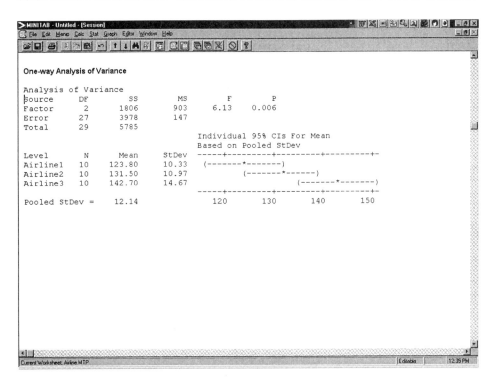

Notice that the test statistic is listed (F = 6.13), as well as the P-value (.006). Since the P-value = .006, and this is smaller than $\alpha = .01$, you should reject the null hypothesis. Thus, there is a difference in the average flight times.

▶ Exercise 1 (pg. 503)   Costs per month of different toothpastes

Open worksheet **Ex10_4-1** found in the MINITAB folder **Chapter 10**. The data for the three different degrees of abrasiveness should be in C1, C2, and C3. To perform a one-way analysis of variance, click on **Stat → ANOVA → One Way (Unstacked)**. Select all three columns and click on **OK**. The results of the test will be in the Session Window.

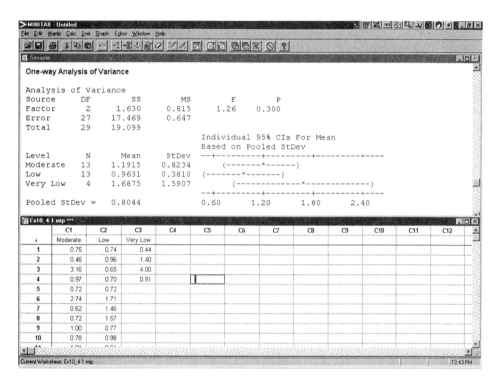

Since the P-value = .300 and is larger than α, you should not reject the null hypothesis. Thus, the average cost per month is the same for all three types of toothpaste.

154 Chapter 10 Chi-Square Tests and the F-Distribution

> **Exercise 5 (pg. 504)** Days spent in a Hospital

Open worksheet **Ex10_4-5** found in the MINITAB folder **Chapter 10**. The data for the four different regions of the United States should be in C1 - C4. To perform a one-way analysis of variance, click on **Stat → ANOVA → One-Way (Unstacked)**. Select all four columns and click on **OK**. The results of the test will be in the Session Window.

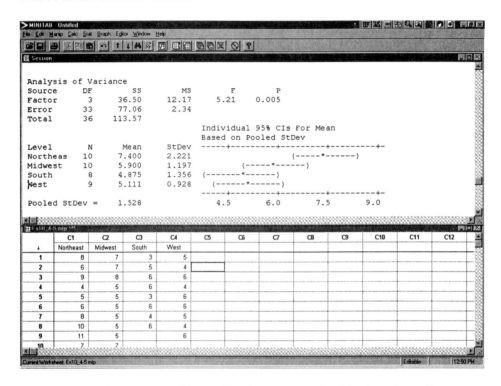

Since the P-value = .005 and is smaller than α, you should reject the null hypothesis. Thus, the average number of days spent in a hospital is not the same for all four regions of the United States.

Technology Lab (pg. 508) Crash Tests

Open worksheet **Tech10_a** found in the MINITAB folder **Chapter 10**. The data for the three types of vehicles should be in C1 - C3.

1. Since the three types of vehicles represent different populations and the three samples were randomly chosen, the samples are independent.

2. For each pair of datasets, perform the Homogeneity of Variance Test. To do this, click on **Manip → Stack/Unstack → Stack Columns.** Select both columns to be stacked on top of each other. **Store stacked data in** C4 and **Store subscripts in** C5. Notice that the subscripts will always be numbers 1 or 2 because in this analysis you are only using two columns at a time. Click on **OK** and look in the Data Window. C4 should have all of the data stored in it and C5 should have the numbers 1 (this tells you that the data value came from the 1st group) or 2 (this tells you that the data value came from the 2nd group) in it. Now, to perform the two-sample F-test, click on **Stat → ANOVA → Homogeneity of Variance.** On the input screen, C4 is the **Response** variable and C5 is the **Factor.** Enter an appropriate **Title** and click on **OK.** This test produces quite a lot of output. However, you are only interested in the results of the F-test.

3. MINITAB has a test for normality. Click on **Stat → Basic Statistics → Normality Test.** Select C1 for the **Variable** and **Kolmogorov-Smirnov** for the **Test of Normality.** Click on **OK** and a normal plot will be displayed.

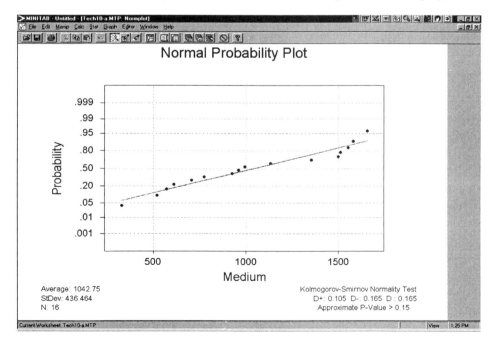

Notice the P-value in the lower right corner of the plot. Since this P-value is larger than .10 (our α), then you can assume that the data is approximately normal. Repeat this test for the other two pairs of columns of data.

4. To do a one-way ANOVA, click on **Stat → ANOVA → One-Way (Unstacked).** Select all three columns and click on **OK.** The results of the test will be in the Session Window. Look at the P-value. If it is smaller than α, you should reject the null hypothesis.

5. Repeat Exercises 1 - 4 using worksheet **Tech10_b** found in the MINITAB folder **Chapter 10.**

Nonparametric Tests

Section 11.1

▶ **Example 3 (pg. 521)** Using the Paired-Sample Sign Test

Open worksheet **Prison** which is found in the **Chapter 11** MINITAB folder. The Before data is in C1 and the After data is in C2. Since we are interested in the difference between C1 and C2, we will create a new column that is C1 - C2. Click on **Calc → Calculator**. **Store result in variable** C3 and calculate the **Expression** C1 - C2. Click on **OK** and C3 should contain the differences. Now perform a 1-sample Sign test on C3. Click on **Stat → Nonparametrics → 1-sample Sign**. Select C3 as the **Variable**. Since you would like to test if the number of repeat offenders has decreased after the special course, you would expect that the differences in C3 would be greater than 0. Thus, enter 0 for **Test median** and select **greater than** for the **Alternative**.

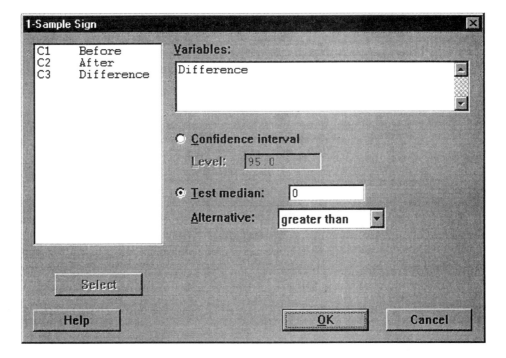

Click on **OK** and the results will be in the Session Window.

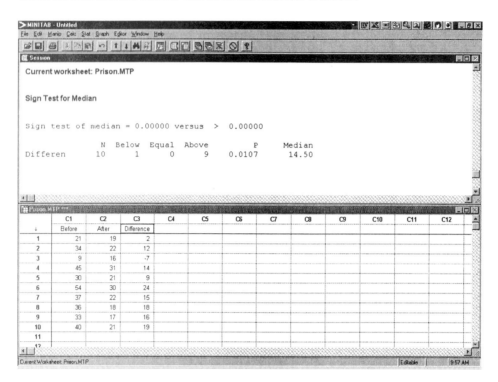

Notice that the P-value = .0107. Since this value is smaller than α=.025, you would reject the null hypothesis. Thus, there is sufficient evidence to conclude that the number of repeat offenders decreases after taking the special course.

▶ **Exercise 3 (pg. 522)** One-sample Sign Test

Enter the data into C1 in the MINITAB Data Window. (Do not type in the $ sign.) To perform the Sign Test, click on **Stat → Nonparametrics → 1-sample Sign**. Select C1 as the **Variable**. Since you would like to test if the median amount of new credit card charges was more than $200, enter 200 for **Test median** and select **greater than** for the **Alternative**.

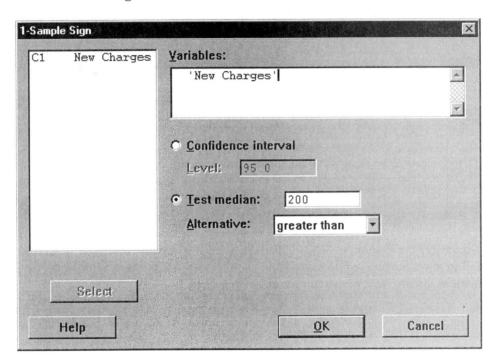

Click on **OK** and the results will be displayed in the Session Window.

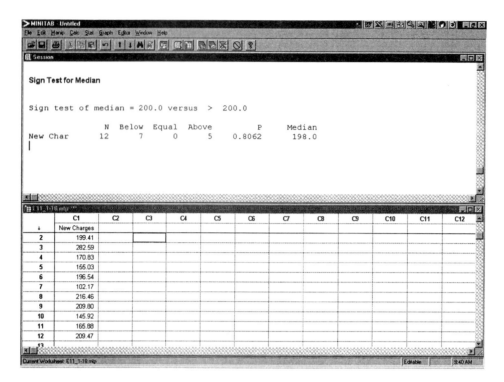

Notice that the P-value = .8062. Since this is such a large P-value, you would fail to reject the null hypothesis. Thus, the accountant can **not** conclude that the median amount of new charges was more than $200.

> **Exercise 15 (pg. 524)** Does therapy reduce headache hours?

Enter Headache hours before therapy into C1 of the MINITAB Data Window and Headache hours after therapy into C2. Calculate the differences. Click on **Calc → Calculator.** Next **Store result in variable** C3 and calculate the **Expression** C1-C2. Click on **OK** and C3 should contain the differences. Now perform a 1-sample Sign test on C3. Click on **Stat → Nonparametrics → 1-sample Sign.** Select C3 as the **Variable.** Since you would like to test if the number of headache hours has decreased after the special therapy, you would expect that the differences in C3 would be greater than 0. Thus, enter 0 for **Test median** and select **greater than** for the **Alternative.** Click on **OK**.

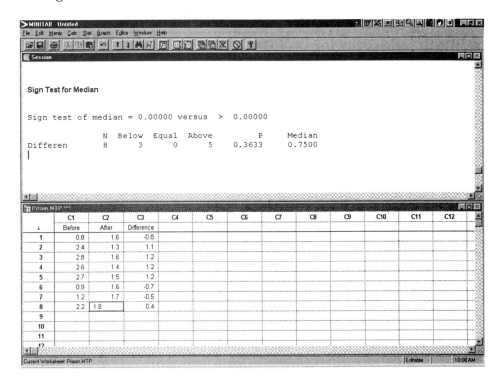

Notice that the P-value = .3633. Since this value is larger than α=.05, you would fail to reject the null hypothesis. Thus, there is not enough evidence to conclude that the number of headache hours decreases with this special therapy.

Section 11.2

> **Example 1 (pg. 528)** Performing a Wilcoxon Signed-Rank Test

Enter the data found on page 528 of the textbook into the MINITAB Data Window. Enter With Music data into C1 and Without Music data into C2. Label each column appropriately. To calculate the differences, click **Calc → Calculator. Store the result in variable** C3 and calculate the **Expression** C1-C2. Click on **OK** and C3 should contain the differences. To perform the Wilcoxon Signed Rank Test, click on **Stat → Nonparametrics → 1-sample Wilcoxon.** You should use C3 for the **Variable.** Since you are using the differences in this example, you want to compare the median difference to 0. So, enter 0 for **Test Median** and choose **not equal** as the **Alternative.**

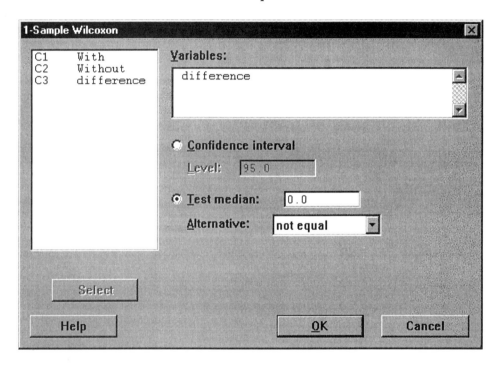

Click on **OK** to view the results of the test in the Session Window.

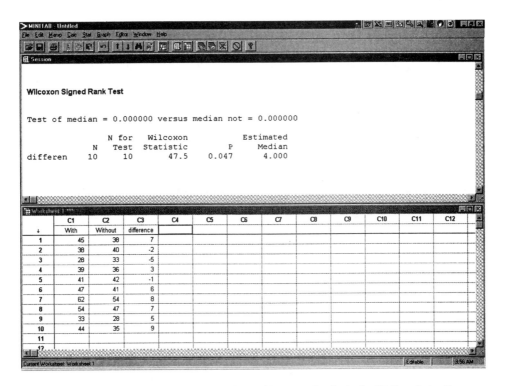

The MINITAB output tells you that the Wilcoxon Statistic is 47.5 and the P-value is .047. Although the textbook tells you to select the smaller of the absolute value of the two sums of the ranks, MINITAB simply uses the sum of the positive ranks. This makes no difference in interpreting the results. The important thing to notice is that the P-value = .047. Since this is smaller than $\alpha = .05$, you would reject the null hypothesis. Thus, there is sufficient evidence to say that music affects the length of the workout sessions.

164 Chapter 11 Nonparametric Tests

▶ Example 2 (pg. 531) Performing a Wilcoxon Rank Sum Test

Open worksheet **earnings** which is found in the **Chapter 11** MINITAB folder. Male earnings are in C1 and Female earnings are in C2. In MINITAB, the Wilcoxon Rank Sum Test is called the Mann-Whitney test. Click on **Stat → Nonparametrics → Mann-Whitney.** Enter C1 for the **First Sample,** C2 for the **Second Sample**, and select n**ot equal** as the **Alternative** since you want to see if there is a difference between the earnings.

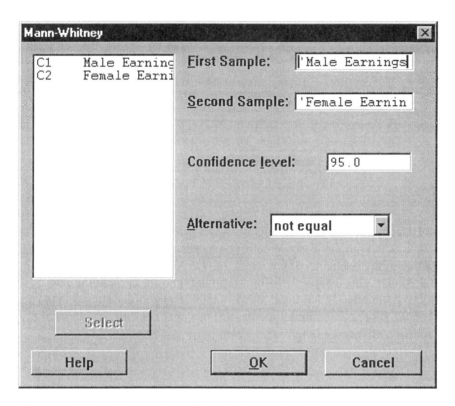

Click on **OK** and the results will be in the Session Window.

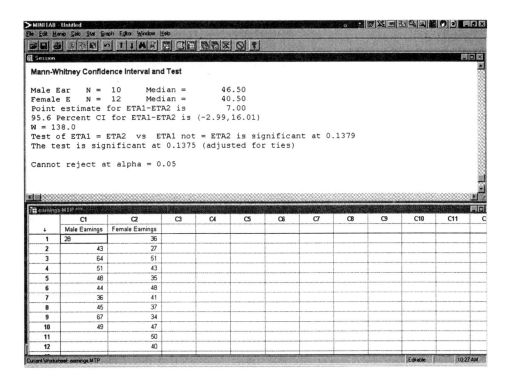

Look at the results carefully. The P-value is .1379. The rank sum for male earnings is also listed, W = 138. In this example, since the P-value is larger than α=.10, you would fail to reject the null hypothesis. Thus, there is no difference between male and female earnings.

Exercise 5 (pg. 534) Teacher salaries

Enter the Ohio salaries into C1 and the Pennsylvania salaries into C2. Click on **Stat → Nonparametrics → Mann-Whitney.** Enter C1 for the **First Sample,** C2 for the **Second Sample**, and select **not equal** as the **Alternative** since you want to see if there is a difference between the salaries in Ohio and Pennsylvania. Click on **OK**. The results should be in the Session Window.

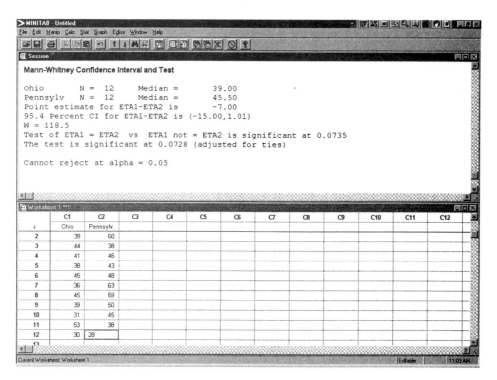

Since the P-value = .0735 and is larger than α=.05, you would fail to reject the null hypothesis. There is not enough evidence to conclude there is a difference in teacher salaries in Ohio and Pennsylvania.

Section 11.3

▶ Example 1 (pg. 539) Performing a Kruskal-Wallis Test

Open worksheet **payrates** which is found in the **Chapter 11** MINITAB folder. The data should be in C1, C2, and C3. To perform a Kruskal-Wallis test, MINITAB requires that the data be stacked into one column with a second column identifying which sample each data value came from. To do this, click on **Manip → Stack/Unstack → Stack Columns.** Select all three columns to be stacked on top of each other. **Store stacked data in** C4 and **Store subscripts in** C5. The subscripts will be numbers 1, 2, or 3 to indicate which column the data value came from.

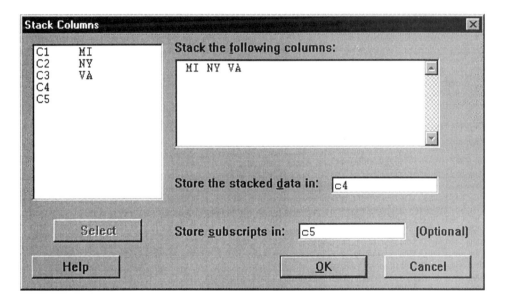

Click on **OK.** Name C4 Payrates and C5 State. Notice that in C5, 1 represents Michigan, 2 represents New York, and 3 represents Virginia.

168 Chapter 11 Nonparametric Tests

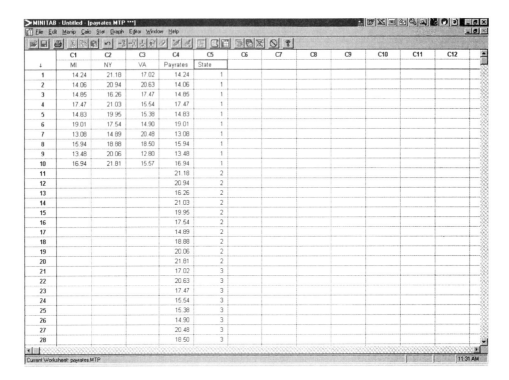

Now, to do the Kruskal-Wallis test, click on **Stat** → **Nonparametrics** → **Kruskal-Wallis**. The **Response** variable is Payrates (C4) and the **Factor** is State (C5).

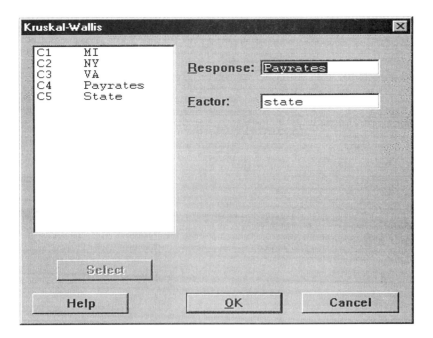

Click on **OK** and the results will be displayed in the Session Window.

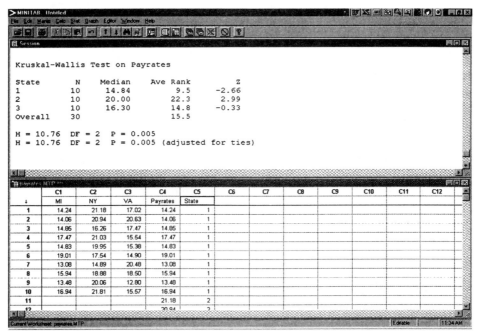

Notice that the test statistic is H=10.76 and the P-value=.005. With such a small P-value, you should reject the null hypothesis. So, there is a difference in the payrates of the three states.

170 Chapter 11 Nonparametric Tests

▶ Exercise 1 (pg. 541) Are the insurance premiums different?

Open worksheet **Ex11_3-1** which is found in the **Chapter 11** MINITAB folder. The data should be in C1, C2, and C3. To perform a Kruskal-Wallis test, MINITAB requires that the data be stacked into one column with a second column identifying which sample each data value came from. To do this, click on **Manip → Stack/Unstack → Stack Columns.** Select all three columns to be stacked on top of each other. **Store stacked data in** C4 and **Store subscripts in** C5. The subscripts will be numbers 1, 2, or 3 to indicate which column the data value came from. Click on **OK.** Name C4 Premiums and C5 State. Notice that in C5, 1 represents California, 2 represents Florida, and 3 represents Illinois. Now, to do the Kruskal-Wallis test, click on **Stat → Nonparametrics → Kruskal-Wallis.** The **Response** variable is Premiums (C4) and the **Factor** is State (C5). Click on **OK** and the results will be displayed in the Session Window.

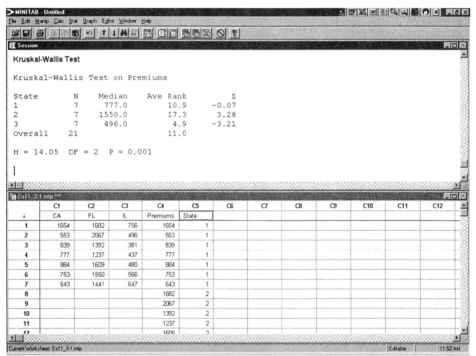

Notice that the test statistic is H=14.05 and the P-value=.001. With such a small P-value, you should reject the null hypothesis. So, there is a difference in the premiums of the three states.

Section 11.4

▶ Example 1 (pg. 544) The Spearman Rank Correlation Coefficient

Enter the data into the MINITAB Data Window. Enter the Beef prices into C1 and the Turkey prices into C2. To rank the data values, click on **Manip → Rank**. On the input screen, you should **Rank data in** C1 and **Store ranks in** C3. When you click on **OK**, the ranks of the Beef prices should be in C3.

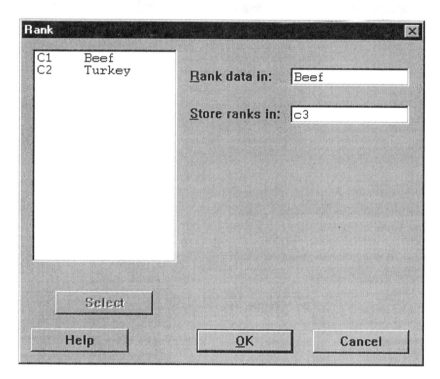

Repeat this using the Turkey prices and storing the ranks in C4. Now you should have the ranks of the data in C3 and C4.

Chapter 11 Nonparametric Tests

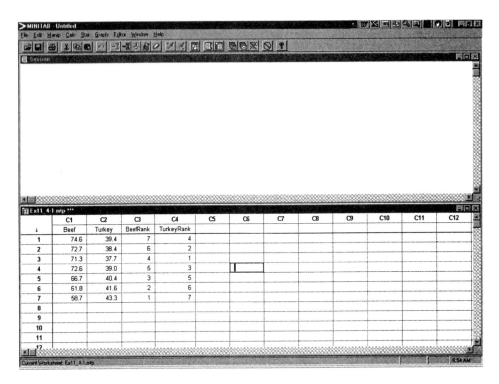

To calculate Spearman's Rank Correlation Coefficient, simply use Pearson's correlation on the ranks of the data. Click on **Stat → Basic Statistics → Correlation.** Enter C3 and C4 for the **Variables** and select **Display p-values.**

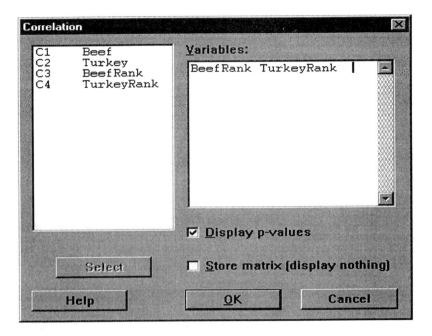

When you click on **OK**, the results will be displayed in the Session Window.

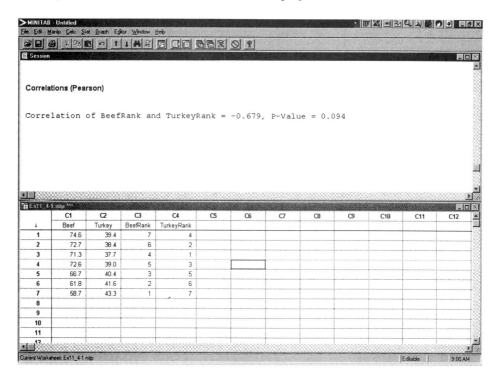

In this example, notice that the Spearman's Rank Correlation Coefficient is -.679 and the P-value is .094. Since this P-value is smaller than $\alpha = .10$, you would reject the null hypothesis. Thus, there is a significant negative correlation between beef and turkey prices from 1990 to 1996.

Exercise 3 (pg. 546) Is Air Conditioner Performance related to Price?

Open worksheet **Ex11_4-3** which is found in the **Chapter 11** MINITAB folder. The overall score is in C1 and the price is in C2. First, rank the data. Click on **Manip → Rank**. On the input screen, you should **Rank data in** C1 and **Store ranks in** C3. When you click on **OK**, the ranks of the Overall Scores should be in C3. Repeat this for the Prices and **store ranks in** C4. Now, calculate the correlation coefficient of the ranks. Click on **Stat → Basic Statistics → Correlation**. Enter C3 and C4 for the **Variables** and select **Display p-values**.

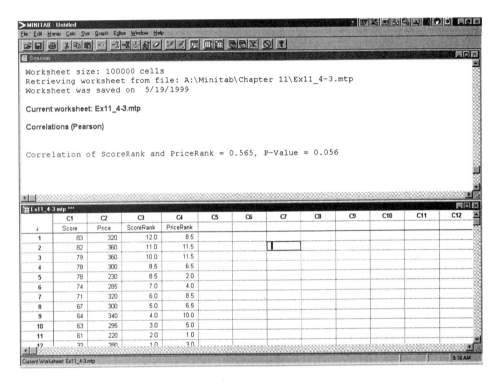

Notice that the Correlation Coefficient is .565 and the P-value is .056. Since the P-value is smaller than α=.10, you would reject the null hypothesis. Thus, there is a significant correlation between Overall Score and Price.

Technology Lab (pg. 549) Selling Prices of Homes

Open worksheet **Tech11_a** which is found in the **Chapter 11** MINITAB folder. The data should be in C1 - C4.

1. Construct a boxplot for all four regions. Click on **Graph → Boxplot.** Enter C1, C2, C3, and C4 (you will have to scroll down to enter C4) for the **Y variables.** Next, since it is easier to compare the data if the boxplots are on the same page, click on **Frame → Multiple graphs.** Select **Overlay graphs on the same page.** Click on **OK.** Click on **Annotation → Title** and enter an appropriate title. Click on **OK.** Finally, click on **Frame → Axis** and enter a **Label,** such as "Thousands of Dollars", for the Y-axis. Click on **OK** twice and compare the boxplots.

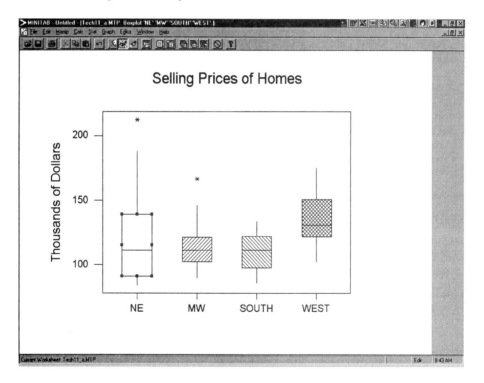

2. To perform a sign test on the data from the South, click on **Stat → Nonparametrics → 1-Sample Sign.** Select South (C3) for the **Variable,** and enter 125 for the **Test Median.** The **Alternative** should be **less than** since the claim is that the median selling price in the South is at least $125,000. Click on **OK** and the results will be displayed in the Session Window.

3. Recall that the Wilcoxon rank sum test is the same as the Mann-Whitney test in MINITAB. Click on **Stat → Nonparametrics → Mann-Whitney.** Enter NE (C1) for the **First sample** and MW (C2) for the **Second sample.** The **Alternative** should be **not equal** since you are testing whether or not the median selling prices are the same. Click on **OK** and the results will be displayed in the Session Window.

4. To perform a Kruskal-Wallis test in MINITAB, the data must be stacked in one column. To do this, click on **Manip → Stack/Unstack → Stack Columns.** Choose all four columns and **store the stacked data in** C5. **Store the subscripts in** C6. When you click on **OK**, C5 and C6 should be filled in. Click on **Stat → Nonparametrics → Kruskal-Wallis.** The **Response** variable is C5 and the **Factor** is C6. Click on **OK** and the results will be displayed in the Session Window.

5. Using the stacked data in C5 and C6, perform a one-way ANOVA. Click on **Stat → ANOVA → One Way.** The **Response** variable is C5 and the **Factor** is C6. Click on **OK** and the results will be displayed in the Session Window.

6. Open worksheet **Tech11_b** which is found in the **Chapter 11** MINITAB folder. Repeat Exercises 1, 3, 4, and 5 using the data in this worksheet.

◄